SpringerBriefs in Neuroscience

For further volumes:
http://www.springer.com/series/8878

Tatiana Borisova

Cholesterol and Presynaptic Glutamate Transport in the Brain

 Springer

Tatiana Borisova
Department of Neurochemistry
Palladin Institute of Biochemistry
National Academy of Sciences of Ukraine
Kiev, Ukraine

ISSN 2191-558X   ISSN 2191-5598 (electronic)
ISBN 978-1-4614-7758-7   ISBN 978-1-4614-7759-4 (eBook)
DOI 10.1007/978-1-4614-7759-4
Springer New York Heidelberg Dordrecht London

Library of Congress Control Number: 2013940268

Printed on acid-free paper

Springer is part of Springer Science+Business Media (www.springer.com)

About the Book

In this monograph, T. Borisova, Ph.D., Dr.Biol.Sci., summarized data of the literature and own results on the role of membrane cholesterol in synaptic transmission, particularly glutamate transport in the presynapse, and also the possibility of neuroprotection by lowering cholesterol was analyzed.

The primary audience for this book is researchers, teachers, graduates, and Ph.D. students of biological and medical specialties, whose activities and interests relate to biochemistry, neurochemistry, neurophysiology, lipid biochemistry, and biotechnology. Highlights of the book may be of interest to a wide range of readers.

Preface

The burden of the modern society is a continuous increase in the number of patients with neurological and neurodegenerative disorders. Nowadays, the analysis of the role of cholesterol in brain functioning is of paramount importance and probably will be so for a long time. Despite recent intense studies on the diverse effects of cholesterol, new regulatory mechanisms are continuously uncovering and the exact role of this steroid in neuronal function and development is further elucidating. Moreover, it was shown recently that cholesterol misbalance has been implicated in the pathogenesis of several neurodegenerative disorders.

The aim of the book was to explore the effects of cholesterol depletion/enrichment of the membranes of the nerve terminals on the key characteristics of glutamatergic neurotransmission. The investigation of the mechanisms of regulation of glutamatergic neurotransmission remains in the mainstream of neurobiological research and is of growing interest till now.

The main question addressed by the author was whether membrane cholesterol modulated glutamate transport in presynaptic nerve terminals, and whether altered cholesterol composition of neuronal membrane could mediate pathogenic mechanisms responsible for neurodegeneration or vice versa have neuroprotective features.

Kiev, Ukraine   Tatiana Borisova

Summary

Glutamate, which is the major excitatory neurotransmitter in the central nervous system, is involved in many aspects of normal brain functioning. It is well-known that disturbances in glutamate transport contribute to neuronal dysfunction as well as the pathogenesis of neurological disorders. Certain level of membrane cholesterol is very important for normal functioning of membrane proteins involved in synaptic transmission.

Cholesterol acceptor methyl-β-cyclodextrin (MβCD) is widely used for effective and selective extraction of membrane cholesterol from a variety of cell types. MβCD (15 mM) reduced the cholesterol content in brain nerve terminals (synaptosomes) by one quarter. The application of MβCD to the synaptosomes and isolated synaptic vesicles led to the gradual leakage of the protons from the vesicles that was shown with pH-sensitive fluorescent dye acridine orange, whereas the application of MβCD complexed with cholesterol (15:0.2) that increased the membrane cholesterol content stimulated additional vesicle acidification. It was suggested that MβCD-mediated cholesterol depletion of the plasma membrane induced the removal of cholesterol from the membrane of synaptic vesicles resulting in immediate dissipation of synaptic vesicle proton gradient, and thereby could provoke redistribution of glutamate between the vesicular and cytosolic pools.

The low level of ambient glutamate is extremely important for proper spontaneous activity and synaptic transmission in the brain. It was examined whether membrane cholesterol modulated the extracellular glutamate level in the nerve terminals and the processes responsible for its maintenance. The ambient glutamate level, being equilibrium between Na$^+$-dependent uptake and tonic release, was increased in rat brain synaptosomes treated with MβCD. This alteration was not due to the change in the activity of glutamine synthetase that was shown with its specific blocker l-methionine sulfoximine. Also, cholesterol-deficient synaptosomes showed a lower initial velocity of glutamate uptake. In the presence of competitive nontransportable inhibitor of glutamate transporters dl-threo-β-benzyloxyaspartate, which significantly reduced glutamate uptake, net tonic release of glutamate from cholesterol-depleted synaptosomes was decreased. It was suggested that cholesterol deficiency altered the intra-to-extracellular glutamate ratio by the reduction of the cytosolic level and the augmentation of the ambient level of the neurotransmitter, thereby favoring a decrease in tonic glutamate release. Thus, increased extracellular concentration of glutamate in cholesterol-deficient nerve terminals was not a result of the changes in tonic release (and/or glutamine synthetase activity), but was set by lack function of glutamate transporters.

In cerebral hypoxia/ischemia, stroke, and traumatic brain injury, the development of neurotoxicity is provoked by enhanced extracellular glutamate, which is released from nerve cells mainly by glutamate transporter reversal—a distinctive feature of these pathological states. Transporter-mediated glutamate release from the synaptosomes: (1) stimulated by depolarization of the plasma membrane; (2) by means of heteroexchange with competitive transportable inhibitor of glutamate transporters dl-threo-β-hydroxyaspartate; (3) in low [Na^+] medium; and (4) during dissipation of the proton gradient of synaptic vesicles by the protonophore carbonyl cyanide-p-trifluoromethoxyphenyl-hydrazon; was decreased under the conditions of cholesterol deficiency. Thus, a decrease in the level of membrane cholesterol attenuated transporter-mediated glutamate release from the nerve terminals. Therefore, lowering cholesterol may be used in neuroprotection in ischemia, stroke, and traumatic brain injury that are associated with an increase in glutamate uptake reversal. This data can explain the neuroprotective effects of statins in these pathological states and provide one of the mechanisms of their neuroprotective action. However, besides these disorders lowering cholesterol may cause harmful consequences decreasing glutamate uptake by the nerve terminals.

Abbreviations

AMPA	α-Amino-3-hydroxy-5-methyl-4-isoxazolepropionic acid
AO	Acridin orange
ATP	Adenosine triphosphate
CNS	Central nervous system
DL-TBOA	dl-Threo-β-benzyloxyaspartate
DL-THA	dl-Threo-β-hydroxyaspartate
EAAC	Excitatory amino acid carrier
EAAT	Excitatory amino acid transporter
EDTA	Ethylenediaminetetraacetic acid
EGTA	Ethylene glycol tetraacetic acid
FCCP	The protonophore carbonyl cyanide-p-trifluoromethoxyphenyl-hydrazon
GDH	Glutamate dehydrogenase
GLAST	l-Glutamate/l-aspartate transporter
Glutamate	l-Glutamic acid
HEPES	4-(2-Hydroxyethyl)-1-piperazineethanesulfonic acid
MβCD	Methyl-β-cyclodextrin
NMDA	N-Methyl-d-aspartic acid
NMDG	N-Methyl-d-glucamine
SDS	Sodium dodecyl sulfate
SNAP	Soluble NSF attachment protein
SNARE (SNAP RECEPTOR)	Soluble N-ethylmaleimide-sensitive fusion protein attachment protein
Tris	Tris(hydroxymethyl)methylamine)
v-SNARE	Vesicle-associated SNAP receptors

Acknowledgments

I am indebted to Dr. N. Krisanova, R. Sivko, and L. Kasatkina for pleasure in joint experimental work and publications, Dr. O. Shatursky for help in BLM experiments, Dr. V. Chernisov and T. Kurchenko for electronic microscopy of synaptosomal preparations.

Contents

Chapter 1
Presynaptic Glutamate Transport in the Brain

Abstract Glutamate is the major excitatory neurotransmitter in the central nervous system, which is involved in many aspects of normal brain functioning, whereas disturbances in glutamate transport contribute to neuronal dysfunction as well as the pathogenesis of neurological disorders.

Glutamate is now recognized as the major excitatory neurotransmitter that essentially mediates all the rapid communications in the central nervous system (CNS). Proper glutamatergic synaptic transmission is essential for basic neuronal communication, synaptic plasticity, and is important for learning and memorizing, attention, control of mood, stress, and anxiety. Abnormal glutamate homeostasis contributes to neuronal dysfunction as well as involves in the pathogenesis of neurological disorders (Danbolt 2001). In the CNS, it is of critical importance to keep low level of extracellular glutamate. From one side, a low extracellular glutamate concentration is normally maintained between episodes of exocytotic release of the neurotransmitter, thereby preventing continual activation of glutamate receptors and protecting neurons from excitotoxic injury (Cavelier and Attwell 2005). Also, low extracellular glutamate ensures a high signal-to-noise ratio for glutamatergic transmission (Jabaudon et al. 1999). From other side, the certain level of ambient glutamate is very important for tonic activation of post- and presynaptic glutamate receptors, e.g., N-methyl-D-aspartate (NMDA) receptors in hippocampal slices, mGluR, α-amino-3-hydroxy-5-methyl-4-isoxazolepropionic acid (AMPA), and kainite receptors (Sah et al. 1989; Dalby and Mody 2003; Cavelier and Attwell 2005), that may be considered as one of the ways for modulation of presynaptic release and regulation of glutamatergic transmission. Recent in vivo microdialysis data revealed that the specific changes in the glutamate concentration in dialysate samples indicated the involvement of ambient glutamate in the information processing of the brain as essential signaling molecules of nonsynaptic transmission (Vizi 2000; Nyitrai et al. 2006).

T. Borisova, *Cholesterol and Presynaptic Glutamate Transport in the Brain*,
SpringerBriefs in Neuroscience 12, DOI 10.1007/978-1-4614-7759-4_1,
© Springer Science+Business Media New York 2013

Neurotransmitter release mainly occurs by synaptic vesicle exocytosis, a very complicated multistep process that involves Ca^{2+}-triggered fusion of docked synaptic vesicles with the plasma membrane. The opening of Ca^{2+} channels in the plasma membrane of the nerve terminals is induced by an action potential. In most synapses, neurotransmitter release is stimulated by Ca^{2+} influx through P/Q- ($Ca_V2.1$) or N-type Ca^{2+} channels ($Ca_V2.2$) (Dietrich et al. 2003), and resulting Ca^{2+} transient stimulates synaptic vesicle exocytosis. After that, synaptic vesicles undergo endocytosis, recycling, and refilling with neurotransmitters and are ready for a new cycle. Nerve terminals are secretory machines dedicated to repeated rounds of release. The interrelation between an action potential and neurotransmitter release from the nerve terminals is regulated by a lot of molecules. Taking into account the above mentioned, it can be concluded that in addition to secretory machines, nerve terminals resemble computational units where the relations of an action potential (input) and neurotransmitter release (output) are continuously changed in response to extra- and intracellular signals (Sudhof 2004). Presynaptic functions, directly or indirectly, involve synaptic vesicles. Synaptic vesicles are small organelles of ~40 nm in diameter, which mediate quantal chemical communication between neurons (Sudhof 2004; Takamori et al. 2006). Synaptic vesicles accumulate neurotransmitters by active transport, and store them at high concentrations. Synaptic vesicles contain two types of obligatory components: transport proteins involved in neurotransmitter uptake and trafficking proteins that participate in synaptic vesicle exo- and endocytosis and recycling. Transport proteins are composed of a V-type proton ATPase that generates the electrochemical gradient, which drives neurotransmitter uptake and neurotransmitter transporters that mediate this uptake. The trafficking proteome of synaptic vesicles includes intrinsic membrane proteins, proteins associated via posttranslational lipid modifications, and peripherally bound proteins (Sudhof 2004). It is generally thought that the size of a synaptic vesicle is determined by the amount of lipid and membrane proteins, which it contains. However, Budzinski et al. (2009) showed recently that isolated synaptic vesicles in solution were reversibly increased in size after filling with glutamate with no apparent addition of lipid or protein molecules. This suggests that synaptic vesicles undergo structural changes as the vesicle fills with the neurotransmitter, and that the presence of the neurotransmitter may be encoded in the vesicle size. Synaptic vesicle protein 2A (SV2A) is required for this phenomenon (Budzinski et al. 2009). It should be noted that synaptic vesicles obtained from the brain did not contain glutamate, due to neurotransmitter leakage during purification in the absence of ATP (Carlson and Ueda 1990).

Membrane fusion is mostly mediated by SNARE proteins (e.g., synaptobrevin, syntaxin, and SNAP25), SM proteins (e.g., Munc 18-1), controlled by synaptotagmins as Ca^{2+} sensors and highly regulated by hundreds of proteins associated with synaptic plasticity (Sudhof et al. 1993; Rothman and Warren 1994; Chen et al. 1999; Chen and Scheller 2001; Wang and Tang 2006). A large number of molecules, which are mainly soluble cytoplasmic proteins, have been suggested to regulate the availability of integral membrane SNARE proteins to form functional neurotransmitter release machinery (Carr and Munson 2007). After opening of fusion pore, synaptic

Fig. 1.1 The ambient extracellular glutamate level

$$[glu]_{extracellular} = [glu]_{tonic\ release} - [glu]_{uptake}$$

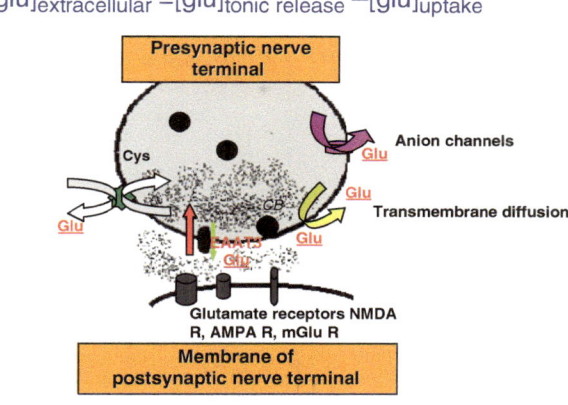

vesicles recycle probably by three alternative pathways: (1) vesicles are reacidified and refilled with neurotransmitters without undocking; (2) vesicles undock and recycle locally to reacidify and refill with neurotransmitters; or (3) vesicles endocytose via clathrin-dependent mechanism and reacidify and refill with neurotransmitters either directly or after passing through an endosomal intermediate (Sudhof 2004).

Great attention has been paid to stimulated release of glutamate by exocytosis, while unstimulated so-called "tonic" or "basal" release (Baker et al. 2002; Bouron 2001; Cavelier et al. 2005; Cavelier and Attwell 2005; Westphalen and Hemmings 2006; Makani and Zagha 2007) is outside of "mainstream" research despite its profound consequences for proper functioning of neurons. Origin of tonic release has not been identified yet, but it is suggested that glutamate enriches the extracellular space mainly by spontaneous exocytosis, through swelling-activated anion channels, cystine–glutamate exchange, and transmembrane diffusion (Rutledge et al. 1998; Jabaudon et al. 1999; Cavelier and Attwell 2005) (Fig. 1.1).

Excess of extracellular glutamate leads to overstimulation of glutamate receptors and causes neurotoxicity. Thus, glutamate acts within the CNS not only as the predominant excitatory neurotransmitter but also as a potent neurotoxin. The balance between physiological and pathological functions of glutamate depends on the rate of removal of the neurotransmitter from the synaptic cleft. The enzymes for glutamate degradation have not been found in the synaptic cleft, so it is the only possibility to maintain a low extracellular glutamate concentration that is realized by high-affinity Na^+-dependent glutamate transporters through neurotransmitter uptake in neurons and glial cells. High-affinity Na^+-dependent glutamate transporters (EAAT 1–5) are plasma membrane proteins with eight putative transmembrane domains, that use Na^+/K^+ electrochemical gradients across the plasma membrane as a driving force (Fonnum 1984; Gegelashvili and Schousboe 1997; Grunewald and Kanner 2000; Danbolt 2001) (Fig. 1.2). In addition to glutamate removal, the glutamate transporters appear to have more sophisticated role in the modulation of neurotransmission. They may modify the time course of synaptic events, the extent of activation, and

Presynaptic nerve terminal

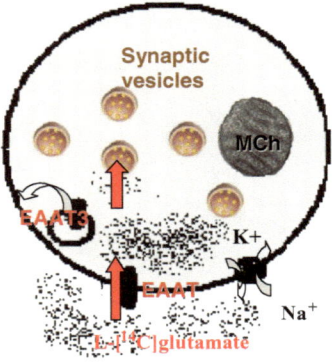

Glutamate uptake activity depends from:

-Na/K electrochemical gradient of the plasma membrane;

-cell surface expression of glutamate transporters (protein kinase C –dependent mechanism of regulation);

-the level of membrane cholesterol;

-acidification of synaptic vesicles.

Fig. 1.2 High-affinity Na^+-dependent glutamate uptake by the nerve terminals

desensitization of receptors outside the synaptic cleft and at neighboring synapses (intersynaptic cross-talk). Also, the glutamate transporters provide the substrate for synthesis of, e.g., GABA, glutathione, and protein, and for energy production (Danbolt 2001). Like glutamate itself, glutamate transporters are involved in almost all aspects of normal and abnormal brain activity (Fonnum 1984; Gegelashvili and Schousboe 1997; Danbolt 2001). The extracellular basal level of the neurotransmitter was determined by a ratio between tonic release and uptake of glutamate.

Translocation of membrane proteins between the plasma membrane and intracellular storage compartments represents one of the important mechanisms of regulation and may contribute to synaptic plasticity (Liu et al. 1999). Such changes in cell-surface expression can up- and down-regulate membrane protein activity within minutes and are much faster than protein synthesis (Danbolt 2001). Recently, it has been shown that activity of glutamate transporters is subjected to substrate-dependent up- and down-regulation mediated by protein kinase C via changes in their cell-surface expression. Using C6 glioma cells expressing EAAC only (Palos et al. 1996; Dowd et al. 1996), it was demonstrated that glutamate transport was rapidly enhanced by phorbol ester-mediated activation of protein kinase C (Dowd and Robinson 1996) that was a result of mobilization of EAAC from intracellular stores (Davis et al. 1998; Robinson 2002). Our preliminary data also showed that the inhibitors of protein kinase C GF 109203X and bisindolylmaleimide II at concentrations of 100 nM decreased glutamate uptake by ~15 %.

However, glutamate transporters can also act in the outward direction, so their function is reversible. Under conditions of energy deprivation and failure of the electrochemical gradient of the plasma membrane, glutamate transporters change the direction of their work and start to release the neurotransmitter into the extracellular space. A decrease in extracellular $[Na^+]$ and/or intracellular $[K^+]$ as well as an increase in extracellular $[K^+]$ and/or intracellular $[Na^+]$ and/or intracellular $[Glu]$ in the nerve terminals thermodynamically favors glutamate transport in the outward direction. In stroke, cerebral hypoxia/ischemia, hypoglycemia, and traumatic brain injury the development of neurotoxicity is provoked by an increase in the concentration of extracellular glutamate. Excessive extracellular glutamate initiating an excessive calcium entry through mainly NMDA ionotropic receptors causes excitotoxicity, neuronal injury, and death. It should be noted that transporter-mediated release is the main mechanism underlying the enhancement of the extracellular glutamate concentration under pathological conditions such as stroke, cerebral hypoxia/ischemia, hypoglycemia, and traumatic brain injury (Jabaudon et al. 1999; Fonnum 1984). Kinetic data predicts that glutamate release through reverse transport can be dramatic because 1 μm^2 of neuronal cell membrane can release 140,000 molecules of glutamate per second at an elevated extracellular $[K^+]$ of 50 mM and reduced $[Na^+]$ of 50 mM ($V_m = -40$ mV) (Grewer et al. 2008). Neuronal glutamate transporters are more reversible in comparison with glial ones. While the neuronal glutamate transporters are functionally converted from an uptake-predominant to a release-predominant state by a reduction in $[Na^+]$ from 145.2 mM to about 60 mM, uptake of glutamate by glial GLT-1 is maintained (Nishida et al. 2004). Therefore, transporter-mediated glutamate release from the neurons mainly contributes to an increase in ambient glutamate concentration under pathological conditions. It is clear that a delay in elevation of ambient glutamate has a potential for preventing brain damage under these pathological states.

Glutamate, which is taken up by the cells, can be used for metabolic purposes, e.g., protein synthesis, energy metabolism, ammonia fixation, or be reused as neurotransmitter (Danbolt 2001). The amino acid glutamine serves an important role as intermediary in both the biosynthesis and metabolism of glutamate (Fonnum 1984). The synthesis of glutamine from glutamate is catalyzed by the cytoplasmic enzyme glutamine synthetase, which was originally found mainly in glial cells (Martinez-Hernandez et al. 1977). This enzyme has been also detected in the nerve terminals but the specific activity of glutamine synthetase is several times lower in neurons in comparison with that observed in astrocytes (Patel et al. 1983; Tansey et al. 1991). Also, glutamine synthesis is important for the detoxification of ammonia because the brain lacks a complete urea cycle. The importance of correct function of glutamine synthetase and the glutamine/glutamate cycle during this detoxifying step in the brain is clear. Glutamine has very low affinity for glutamate receptors and does not interfere with synaptic signaling even at the high concentration (0.3 mM), at which it normally occurs in the brain extracellular fluid (Erecińska and Silver 1990). Most studies on glutamate–glutamine interrelationships have used synaptosomes, acutely prepared slices, or dispersed cell cultures in conjunction with biochemical assays (Laake et al. 1995).

Glutaminase catalyzes the hydrolysis of the amide group of glutamine and forms glutamate and ammonia (Kvamme et al. 2001). In contrast to glutamine synthetase, the activity and expression of this enzyme is higher in neurons than in astrocytes (Hogstad et al. 1988).

Glutamate dehydrogenase is located in the mitochondria although there is evidence for extramitochondrial localization of this enzyme (Mastorodemos et al. 2005).

Glutamate dehydrogenase catalyzes the reaction of oxidative deamination between glutamate, α-ketoglutarate, and ammonia with NAD or NADP as the coenzyme. However, a high content of ammonia in discrete microenvironments of the mitochondria may be found in glutamatergic neurons, where the reaction may reverse to reductive amination, i.e., formation of glutamate from α-ketoglutarate and ammonia (Bak et al. 2005, 2006; Rowley et al. 2012).

Taking into consideration that disturbances in glutamate transport contribute to neuronal dysfunction and the pathogenesis of neurological disorders, it is clear that glutamate transporters and the enzymes involved in glutamate metabolism are perspective targets for therapy development. Unfortunately, the efforts to exploit the mechanisms through which glutamate is synthesized, released, and removed from the synaptic cleft have failed to lead to the development of any commercially available neuroprotective or antiepileptic therapies (Rowley et al. 2012).

References

Bak LK, Sickmann HM, Schousboe A, Waagepetersen HS (2005) Activity of the lactate-alanine shuttle is independent of glutamate-glutamine cycle activity in cerebellar neuronal-astrocytic cultures. J Neurosci Res 79:88–96

Bak LK, Schousboe A, Waagepetersen HS (2006) The glutamate/GABA-glutamine cycle: aspects of transport, neurotransmitter homeostasis and ammonia transfer. J Neurochem 98:641–653

Baker DA, Xi ZX, Shen H et al (2002) The origin and neuronal function of in vivo nonsynaptic glutamate. J Neurosci 22:9134–9141

Bouron A (2001) Modulation of spontaneous quantal release of neurotransmitters in the hippocampus. Prog Neurobiol 63:613–635

Budzinski KL, Allen RW, Fujimoto BS, Kensel-Hammes P, Belnap DM, Bajjalieh SM, Chiu DT (2009) Large structural change in isolated synaptic vesicles upon loading with neurotransmitter. Biophys J 97:2577–2584

Carlson MD, Ueda T (1990) Accumulated glutamate levels in the synaptic vesicle are not maintained in the absence of active transport. Neurosci Lett 110:325–330

Carr CM, Munson M (2007) Tag team action at the synapse. EMBO Rep 89:834–838

Cavelier P, Attwell D (2005) Tonic release of glutamate by a DIDS-sensitive mechanism in rat hippocampal slices. J Physiol 564:397–410

Cavelier P, Hamann M, Rossi D et al (2005) Tonic excitation and inhibition of neurons: ambient transmitter sources and computational consequences. Prog Biophys Mol Biol 87:3–16

Chen YA, Scheller RH (2001) SNARE-mediated membrane fusion. Nat Rev Mol Cell Biol 2(2):98–106

Chen YA, Scales SJ, Patel SM, Doung YC, Scheller RH (1999) SNARE complex formation is triggered by Ca^{2+} and drives membrane fusion. Cell 97:165–174

Dalby NO, Mody I (2003) Activation of NMDA receptors in rat dentate gyrus granule cells by spontaneous and evoked transmitter release. J Neurophysiol 90:786–797

Danbolt NC (2001) Glutamate uptake. Prog Neurobiol 65:1–105

Davis KE, Straff DJ, Weinstein EA, Bannerman PG, Correale DM, Rothstein JD, Robinson MB (1998) Multiple signaling pathways regulate cell surface expression and activity of the excitatory amino acid carrier 1 subtype of Glu transporter in C6 glioma. J Neurosci 18:2475–2485

Dietrich D, Kirschstein T, Kukley M, Pereverzev A, von der Brelie C, Schneider T, Beck H (2003) Functional specialization of presynaptic Cav2.3 Ca^{2+} channels. Neuron 39:483–496

Dowd LA, Robinson MB (1996) Rapid stimulation of EAAC1-mediated Na^+-dependent L-glutamate transport activity in C6 glioma cells by phorbol ester. J Neurochem 67:508–516

Dowd LA, Coyle AJ, Rothstein JD, Pritchett DB, Robinson MB (1996) Comparison of Na$^+$-dependent glutamate transport activity in synaptosomes, C6 glioma, and Xenopus oocytes expressing excitatory amino acid carrier 1 (EAAC1). Mol Pharmacol 49(3):465–473

Erecińska M, Silver IA (1990) Metabolism and role of glutamate in mammalian brain. Prog Neurobiol 35(4):245–296

Fonnum F (1984) Glutamate: a neurotransmitter in mammalian brain. J Neurochem 42:1–11

Gegelashvili G, Schousboe A (1997) High affinity glutamate transporters: regulation of expression and activity. Mol Pharmacol 52:6–15

Grewer C, Gameiro A, Zhang Z, Tao Z, Braams S, Rauen T (2008) Glutamate forward and reverse transport: from molecular mechanism to transporter-mediated release after ischemia. IUBMB Life 60:609–619

Grunewald M, Kanner BI (2000) The accessibility of a novel reentrant loop of the glutamate transporter GLT-1 is restricted by its substrate. J Biol Chem 275:9684–9689

Hogstad S, Svenneby G, Torgner IA, Kvamme E, Hertz L, Schousboe A (1988) Glutaminase in neurons and astrocytes cultured from mouse brain: kinetic properties and effects of phosphate, glutamate, and ammonia. Neurochem Res 13:383–388

Jabaudon D, Shimamoto K, Yasuda-Kamatani Y (1999) Inhibition of uptake unmasks rapid extracellular turnover of glutamate of nonvesicular origin. Proc Natl Acad Sci USA 96:8733–8738

Kvamme E, Torgner IA, Roberg B (2001) Kinetics and localization of brain phosphate activated glutaminase. J Neurosci Res 66:951–958

Laake JH, Slyngstad TA, Haug FMS, Ottersen OP (1995) Glutamine from glial cells is essential for the maintenance of the nerve terminal pool of glutamate: immunogold evidence from hippocampal slice cultures. J Neurochem 65:871–881

Liu Y, Krantz DE, Waites C, Edwards RH (1999) Membrane trafficking of neurotransmitter transporters in the regulation of synaptic transmission. Trends Cell Biol 9:356–363

Makani S, Zagha E (2007) Out of the cleft: the source and target of extra-synaptic glutamate in the CA1 region the hippocampus. J Physiol 582(2):479–480

Martinez-Hernandez A, Bell KP, Norenberg MD (1977) Glutamine synthetase: glial localization in brain. Science 195(4284):1356–1358

Mastorodemos V, Zaganas I, Spanaki C, Bessa M, Plaitakis A (2005) Molecular basis of human glutamate dehydrogenase regulation under changing energy demands. J Neurosci Res 79:65–73

Nishida A, Iwata H, Kudo Y et al (2004) Measurement of glutamate uptake and reversed transport by rat synaptosome transporters. Biol Pharm Bull 27:813–816

Nyitrai G, Kékesi KA, Juhász G (2006) Extracellular level of GABA and Glu: in vivo microdialysis-HPLC measurements. Curr Top Med Chem 6:935–940

Palos TP, Ramachandran B, Boado R, Howard BD (1996) Rat C6 and human astrocytic tumor cells express a neuronal type of glutamate transporter. Mol Brain Res 37:297–303

Patel AJ, Hunt A, Tahourdin CSM (1983) Regulation of in vivo glutamine synthetase activity by glucocorticoids in the developing rat brain. Dev Brain Res 10:83–91

Robinson MB (2002) Regulated trafficking of neurotransmitter transporters: common notes but different melodies. J Neurochem 80:1–11

Rothman JE, Warren G (1994) Implications of the SNARE hypothesis for intracellular membrane topology and dynamics. Curr Biol 4:220–233

Rowley NM, Madsen KK, Schousboe A, White HS (2012) Glutamate and GABA synthesis, release, transport and metabolism as targets for seizure control. Neurochem Int 61:546–558

Rutledge EM, Aschner M, Kimelberg HK (1998) Pharmacological characterization of swelling-induced D-[3H]aspartate release from primary astrocyte cultures. Am J Physiol 274:1511–1520

Sah P, Hestrin S, Nicoll RA (1989) Tonic activation of NMDA receptors by ambient glutamate enhances excitability of neurons. Science 246:815–818

Sudhof TC (2004) The synaptic vesicle cycle. Annu Rev Neurosci 27:509–547

Sudhof TC, De Camilli P, Niemann H, Jahn R (1993) Membrane fusion machinery: insights from synaptic proteins. Cell 75:1–4

Takamori S, Holt M, Stenius K, Lemke EA, Gronborg M, Riedel D, Urlaub H, Schenck S, Brügger
    B, Ringler P, Müller SA, Rammner B, Gräter F, Hub JS, De Groot BL, Mieskes G, Moriyama Y,
    Klingauf J, Grubmüller H, Heuser J, Wieland F, Jahn R (2006) Molecular anatomy of a trafficking
    organelle. Cell 127:831–846
Tansey FA, Farooq M, Cammer W (1991) Glutamine synthetase in oligodendrocytes and astrocytes:
    new biochemical and immunocytochemical evidence. J Neurochem 56:266–272
Vizi ES (2000) Role of high-affinity receptors and membrane transporters in nonsynaptic
    communication and drug action in the central nervous system. Pharmacol Rev 52:63–89
Wang Y, Tang BL (2006) SNAREs in neurons-beyond synaptic vesicle exocytosis. Mol Membr
    Biol 23(5):377–384
Westphalen RI, Hemmings HC Jr (2006) Volatile anesthetic effects on glutamate versus GABA
    release from isolated rat cortical nerve terminals: basal release. J Pharmacol Exp Ther
    316:208–215

Chapter 2
Cholesterol and Its Role in Synaptic Transmission

Abstract Certain level of membrane cholesterol, which is an abundant constituent of eukaryotic membranes, is very important for normal functioning of a number of membrane proteins involved in synaptic transmission, such as ion channels, pumps, receptors, and transporters, while the alterations in cholesterol content change the property of membranes and the activity of these proteins. Moreover, cholesterol deficiency has been implicated in the pathogenesis of several neurodegenerative disorders.

Cholesterol is required to establish proper membrane permeability, fluidity, phase behavior, and thickness. Cholesterol stabilizes membranes and provides order to membranes, which are in the fluid phase, vice versa to membranes in the gel phase. Cholesterol regulates lipid chain order underlying many of the properties and functions of membrane including permeability to water and other molecules. Since cholesterol creates tighter packing of the membrane, it reduces movement of permeants (Kroes and Ostwald 1971; Crockett 1998). Many properties of biological membranes are particularly influenced by temperature. It is membrane order (increasing temperatures fluidize membranes, however decreasing temperatures order membranes) and also membrane phase transition, permeability, and thickness. Since cholesterol has pronounced effects on the physical properties of membranes, it may be used to minimize effects of temperature on membrane structure and function. Modulation of the cholesterol level could reverse, or at least ameliorate, temperature-induced perturbations in the properties of membranes (Crockett 1998).

In recent years, much attention has been focused on the role of membrane cholesterol in the regulation of the cellular functions (Subtil et al. 1999; Launikonis and Stephenson 2001; Mauch et al. 2001; Hill et al. 2002; Hering et al. 2003; Pfrieger 2003; Churchward et al. 2005; Lange et al. 2005; Rohrbough and Broadie 2005; Allen et al. 2007; Cho et al. 2007; Wasser et al. 2007). Membrane microdomains enriched with cholesterol are considered to serve as scaffolding regions, where the interface of different signal transduction pathways occurs (Martens et al. 2004).

T. Borisova, *Cholesterol and Presynaptic Glutamate Transport in the Brain*, SpringerBriefs in Neuroscience 12, DOI 10.1007/978-1-4614-7759-4_2, © Springer Science+Business Media New York 2013

The CNS, which is equal to 2 % of body mass, keeps a special place among other systems of the organism, because it contains approximately a quarter of total unesterified cholesterol (Dietschy and Turley 2001; Chattopadhyay and Paila 2007). It is necessary to underline that cholesterol is a prominent component of both partners of exocytotic process, i.e., the presynaptic plasma membrane and synaptic vesicles (Deutsch and Kelly 1981). The level of cholesterol determines the fluidity and curvature of the vesicle membranes (Cevc and Richardsen 1999; Zamir and Charlton 2006) and is important for the formation of the complex between synaptophysin, a major cholesterol-binding protein in brain synaptic vesicles, and synaptobrevin (Thiele et al. 2000), thereby affecting synaptic efficiency (Mitter et al. 2003).

Exocytosis, release mediated by transporter reversal, and unstimulated release and uptake of glutamate, which are tightly associated with membrane components of the cells, are not constant and subjected to regulation at multiple levels including modulation of lipid composition of the plasma membrane. Certain level of membrane cholesterol is very important for normal functioning of a number of membrane proteins involved in synaptic transmission, such as ion channels, pumps, receptors, and transporters, while the alterations in cholesterol content change the fusibility of membranes and the activity of these proteins (Fong and McNamee 1986; Jennings et al. 1999; Burger et al. 2000; Lang et al. 2001; Sooksawate and Simmonds 2001; Romanenko et al. 2002; Chou et al. 2003; Eroglu et al. 2003; Kato et al. 2003; Taverna et al. 2004; Butchbach et al. 2004; Xia et al. 2004, 2007; Dalskov et al. 2005; González et al. 2007; Zidovetzki and Levitan 2007). Localization and trafficking/internalization of neurotransmitter receptors (Fong and McNamee 1986; Burger et al. 2000; Sooksawate and Simmonds 2001; Eroglu et al. 2003) and activity of specific plasma membrane transporters (Butchbach et al. 2004; González et al. 2007) are greatly influenced by cholesterol depletion. Cholesterol is also important for the maintenance of synapse organization, processes of synaptogenesis, and synaptic vesicles recycling (Mauch et al. 2001; Hering et al. 2003; Pfrieger 2003; Salaun et al. 2004; Rohrbough and Broadie 2005; Wasser et al. 2007).

It was shown recently that cholesterol depletion differently influenced transporter-mediated glutamate uptake in mouse brain plasma membrane vesicles, primary cortical cultures, and in hippocampal astrocytes. The latter was analyzed by Tsai et al. (2006), who demonstrated that cholesterol deprivation increased uptake of a stable analogue of glutamate, D-[^{3}H]aspartate, by glutamate transporters in hippocampal astrocyte cultures. In contrast, Butchbach et al. (2004) revealed a significant decrease in glutamate transporter activity in cholesterol-depleted primary cortical cultures as well as mouse brain plasma membrane vesicles. It was suggested that after extraction of membrane cholesterol the alterations in the functioning of glutamate transporters, which are organized in a clustered manner on the plasma membrane of neurons, might be caused by the reduction of cluster number and cluster size and also could originate from changed cell surface expression of glutamate transporters. Thus, in nerve terminals, the effect of cholesterol depletion on glutamate uptake mediated by glutamate transporters was opposite to that observed in astrocytes. Glutamate transporters of both cells more or less contribute to the maintenance of the low level of ambient glutamate (Danbolt 2001; Brasnjo an Otis 2001;

Cavelier and Attwell 2005; Beurrier et al. 2009), so the alteration in membrane cholesterol content may cause diverse changes in this level in neurons and astrocytes. To uncover different underlying mechanisms, the effects of cholesterol depletion should be examined independently in each type of the cells. Despite recent intense research and numerous experimental data on the diverse effects of cholesterol, new regulatory mechanisms are continuously uncovering and the exact role of this steroid in neuronal function and development is further elucidating (Subtil et al. 1999; Launikonis and Stephenson 2001; Mauch et al. 2001; Hill et al. 2002; Hering et al. 2003; Pfrieger 2003; Salaun et al. 2004; Rohrbough and Broadie 2005; Lange et al. 2005; Churchward et al. 2005; Salaun et al. 2005; Allen et al. 2007; Chattopadhyay and Paila 2007; Cho et al. 2007; Wasser et al. 2007).

Cholesterol-depleted brain synaptosomes, plasma membrane vesicles, rat primary cortical cultures, and cell lines are routinely used to evaluate the physiological role of cholesterol in neuronal function. The treatment with cyclodextrins, which are a family of cyclic oligosaccharides composed of a lipophilic cavity and hydrophilic outer surface, is the common methodological approach in this research. Methyl-β-cyclodextrin (MβCD), which contains seven a-(1,4) linked glycosyl units, is known as an effective cholesterol-depleting agent (Yancey et al. 1996; Rodal et al. 1999; Jadot et al. 2001; Steck et al. 2002; Singh et al. 2002; Barnes et al. 2004). It should be noted that the measurements were carried out in the presence of the acceptor in the incubation media (Yancey et al. 1996; Jadot et al. 2001; Butchbach et al. 2004; Zamir and Charlton 2006) as well as after washing of MβCD (Levitan et al. 2000; Lang et al. 2001; Taverna et al. 2004; Wasser et al. 2007). MβCD evokes cholesterol from the plasma membrane with the half-time of cholesterol transfer across the cell bilayer equaled to approximately 1 s at 37 °C (Steck et al. 2002). Biexponential kinetics of cyclodextrin-mediated cholesterol efflux was shown by Yancey et al. (1996) indicating the existence of two different pools of cholesterol in cells: a fast and slow pool with a half-time of 19–23 s and 15–30 min, respectively. It should be underlined that the fast pool of cholesterol was completely restored after 40-min recovery period (Yancey et al. 1996). Experimental data showed that the treatment of cells with MβCD not only resulted in cholesterol removal from the plasma membrane but also induced more dramatic changes in the level of cholesterol in intracellular membranes, such as lysosomal membrane and membrane of endoplasmic reticulum (Jadot et al. 2001; Lange et al. 2005; Zidovetzki and Levitan 2007). In neurochemical studies, it should be kept in mind that cholesterol is a prominent component of synaptic vesicle membrane (Deutsch and Kelly 1981). Since MβCD-mediated cholesterol extraction from the plasma membrane was a complicated multistep process followed by the redistribution and restoration of steroid (Yancey et al. 1996; Jadot et al. 2001; Steck et al. 2002; Lange et al. 2004; Zidovetzki and Levitan 2007), it may be suggested that the cellular mechanisms underlying the changes in the key processes of synaptic transmission could differ in dependence of the manner of the acceptor application. It is still unclear exactly what intracellular processes occur just after the addition of MβCD to the cells, i.e., during acute depletion of membrane cholesterol.

In aqueous solutions, cyclodextrin molecules self-assemble and form nano-sized aggregates. Based on experimental data, Miyajima et al. (1983, 1986) showed the spontaneous formation of aggregates of cyclodextrins in aqueous medium. The anomalously low solubility of β-cyclodextrins in aqueous solutions may be explained by intensive formation of aggregates, which became notable at concentrations of higher than 3 mM (Bonini et al. 2006; Messner et al. 2010). This fact creates the potential to develop sophisticated drug delivery systems with cyclodextrins (Messner et al. 2010). It was demonstrated that the aggregate formation was concentration-dependent process, which was increased with the enhancement of cyclodextrin concentration. The aggregation of cyclodextrin complexes can be driven by hydrophobic guests, especially by those guests that tend to self-aggregate themselves. Formation of guest inclusion complexes can increase significantly the tendency of cyclodextrin molecules to form aggregates. This is especially critical during drug formulation studies where concentrations of both drug and cyclodextrins are relatively high (Messner et al. 2010).

Taking into account the above data, it is suggested that cholesterol can be a potent endogenous modulator of glutamate transport in the brain. It is important because there are no substances used clinically to modulate functioning of glutamate transporters in the brain. The main questions should be asked, "How altered cholesterol composition of neuronal membrane could mediate pathogenic mechanisms responsible for the neurodegeneration and whether reduced cholesterol content of neuronal membrane can modulate pathogenic mechanisms underlying the development of neurotoxicity?"

References

Allen JA, Halverson-Tamboli RA, Rasenick MM (2007) Lipid rafts microdomains and neurotransmitter signaling. Nat Rev Neurosci 8:128–140

Barnes K, Ingram JC, Bennett MDM et al (2004) Methyl-beta-cyclodextrin stimulates glucose uptake in Clone 9 cells: a possible role for lipid rafts. Biochem J 378:343–351

Beurrier C, Bonvento G, Kerkerian-Le Goff L, Gubellini P (2009) Role of glutamate transporters in corticostriatal synaptic transmission. Neuroscience 158:1608–1615

Bonini M, Rossi S, Karlsson G, Almgren M, Lo Nostro P, Baglioni P (2006) Self-assembly of beta-cyclodextrin in water. Part 1: Cryo-TEM and dynamic and static light scattering. Langmuir 22:1478–1484

Brasnjo G, Otis T (2001) Neuronal glutamate transporters control activation of postsynaptic metabotropic glutamate receptor and influence cerebellar long-term depression. Neuron 31:607–616

Burger K, Gimpl G, Fahrenholz F (2000) Regulation of receptor function by cholesterol. Cell Mol Life Sci 57:1577–1592

Butchbach M, Tian G, Guo H, Lin CG (2004) Association of excitatory amino acid transporters, especially EAAT2, with cholesterol-rich lipid raft microdomains. J Biol Chem 279: 34388–34396

Cavelier P, Attwell D (2005) Tonic release of glutamate by a DIDS-sensitive mechanism in rat hippocampal slices. J Physiol 564:397–410

Cevc G, Richardsen H (1999) Lipid vesicles and membrane fusion. Adv Drug Deliv Rev 38:207–232

Chattopadhyay A, Paila YD (2007) Lipid-protein interactions, regulation and dysfunction of brain cholesterol. Biochem Biophys Res Commun 354:627–633

Cho WJ, Jeremic A, Jin H et al (2007) Neuronal fusion pore assembly requires membrane cholesterol. Cell Biol Int 31:1301–1308

Chou YC, Lin SB, Tsai LH, Tsai HI, Lin CM (2003) Cholesterol deficiency increases the vulnerability of hippocampal glia in primary culture to glutamate-induced excitotoxicity. Neurochem Int 43:197–209

Churchward MA, Rogasevskaia T, Hofgen J et al (2005) Cholesterol facilitates the native mechanism of Ca^{2+}-triggerted membrane fusion. J Cell Sci 118:4833–4848

Crockett EL (1998) Cholesterol function in plasma membranes from ectotherms: membrane-specific roles in adaptation to temperature. Am Zool 38:291–304

Danbolt NC (2001) Glutamate uptake. Prog Neurobiol 65:1–105

Dalskov SM, Immerdal L, Niels-Christiansen LL, Hansen GH, Schousboe A, Danielsen EM (2005) Lipid raft localization of GABA A receptor and Na^+, K^+-ATPase in discrete microdomain clusters in rat cerebellar granule cells. Neurochem Int 46:489–499

Deutsch JW, Kelly RB (1981) Lipids of synaptic vesicles: relevance to the mechanism of membrane fusion. Biochemistry 20:378–385

Dietschy JM, Turley SD (2001) Cholesterol metabolism in the brain. Curr Opin Lipidol 12:105–112

Eroglu C, Bruger B, Wieland F et al (2003) Glutamate-binding affinity of Drosophila metabotropic glutamate receptor is modulated by association with lipid rafts. Proc Natl Acad Sci USA 100:10219–10224

Fong TM, McNamee MG (1986) Correlation between acetylcholine receptor function and structural properties of membranes. Biochemistry 25:830–840

González MI, Susarla BT, Fournier KM (2007) Constitutive endocytosis and recycling of the neuronal glutamate transporter, excitatory amino acid carrier 1. J Neurochem 103:1917–1931

Hering H, Lin CC, Sheng M (2003) Lipid rafts in the maintenance of synapses, dendritic spines, and surface AMPA receptor stability. J Neurosci 23:3262–3271

Hill W, An B, Johnson J (2002) Endogenously expressed epithelial sodium channel is present in lipid rafts in A6 cells. J Biol Chem 277:33541–33544

Jadot M, Andrianaivo F, Dubois F, Wattiaux R (2001) Effects of methylcyclodextrin on lysosomes. Eur J Biochem 268:1392–1399

Jennings LJ, Xu QW, Firth TA (1999) Cholesterol inhibits spontaneous action potentials and calcium currents in guinea pig gallbladder smooth muscle. Am J Physiol 277:1017–1026

Kato N, Nakanishi M, Hirashima N (2003) Cholesterol depletion inhibits store-operated calcium currents and exocytotic membrane fusion in RBL-2H3 cells. Biochemistry 42:11808–11814

Kroes J, Ostwald R (1971) Erythrocyte membranes—effect of increased cholesterol content on permeability. Biochim Biophys Acta 249:647–650

Lang T, Bruns D, Wenzel D et al (2001) SNAREs are concentrated in cholesterol-dependent clusters that define docking and fusion sites for exocytosis. EMBO J 20:2202–2213

Lange Y, Ye J, Steck TL (2004) How cholesterol homeostasis is regulated by plasma membrane cholesterol in excess of phospholipids. Proc Natl Acad Sci USA 101(32):11664–11667

Lange Y, Ye J, Steck TL (2005) Activation of membrane cholesterol by displacement from phospholipids. J Biol Chem 280:36126–36131

Launikonis BS, Stephenson DG (2001) Effects of membrane cholesterol manipulation on excitation-contraction coupling in skeletal muscle of the toad. J Physiol 534:71–85

Levitan I, Christian AE, Tulenko TN, Rothblat GH (2000) Membrane cholesterol content modulates activation of volume-regulated anion current in bovine endothelial cells. J Gen Physiol 115:405–416

Martens J, O'Connell K, Tamkun M (2004) Targeting of ion channels to membrane microdomains: localization of Kv channels to lipid rafts. Trends Pharmacol Sci 25:16–21

Mauch DH, Nägler K, Schumacher S et al (2001) CNS synaptogenesis promoted by glia-derived cholesterol. Science 294:1354–1357

Messner M, Kurkov SV, Jansook P, Loftsson T (2010) Self-assembled cyclodextrin aggregates and nanoparticles. Int J Pharm 387:199–208

Mitter D, Reisinger C, Hinz B (2003) The synaptophysin/synaptobrevin interaction critically depends on the cholesterol content. J Neurochem 84:35–42

Miyajima K, Sawada M, Nakagaki M (1983) Viscosity B-coefficients, apparent molar volumes, and activity-coefficients for alpha-cyclodextrin and gamma-cyclodextrin in aqueous-solutions. Bull Chem Soc Jpn 56:3556–3560

Miyajima K, Mukai T, Nakagaki M, Otagiri M, Uekama K (1986) Activity-coefficients of dimethyl-beta-cyclodextrin in aqueous-solutions. Bull Chem Soc Jpn 59:643–644

Pfrieger FW (2003) Cholesterol homeostasis and function in neurons of the central nervous system. Cell Mol Life Sci 60:1158–1171

Rodal SK, Skretting G, Garred O, Vilhardt F, van Deurs B, Sandvig K (1999) Extraction of cholesterol with methyl-beta-cyclodextrin perturbs formation of clathrin-coated endocytic vesicles. Mol Biol Cell 10:961–974

Rohrbough J, Broadie K (2005) Lipid regulation of the synaptic vesicle cycle. Nat Rev Neurosci 6:139–150

Romanenko VG, Rothblat GH, Levitan I (2002) Modulation of endothelial inward-rectifier K+ current by optical isomers of cholesterol. Biophys J 83:3211–3222

Salaun C, James DJ, Chamberlain LH (2004) Lipid rafts and the regulation of exocytosis. Traffic 5:1–10

Salaun C, Gould GW, Chamberlain LH (2005) The SNARE proteins SNAP-25 and SNAP-23 display different affinities for lipid rafts in PC12 cells. Regulation by distinct cysteine-rich domains. J Biol Chem 280(2):1236–1240

Singh M, Sherma R, Banerjee U (2002) Biotechnological application of cyclodextrins. Biotechnol Adv 20:341–359

Sooksawate T, Simmonds MA (2001) Effects of membrane cholesterol on the sensitivity of the GABA(A) receptor to GABA in acutely dissociated rat hippocampal neurones. Neuropharmacology 40:178–184

Steck TL, Ye J, Lange Y (2002) Probing red cell membrane cholesterol movement with cyclopdextrin. Biophys J 83:2118–2125

Subtil A, Gaidarov I, Kobylarz K et al (1999) Acute cholesterol depletion inhibits clathrin-coated pit budding. Proc Natl Acad Sci USA 96:6775–6780

Taverna E, Saba E, Rowe J et al (2004) Role of lipid microdomains in P/Q-type calcium channel (Cav2.1) clustering and function in presynaptic membranes. J Biol Chem 279:5127–5134

Thiele C, Hannah MJ, Fahrenholz F, Huttner WB (2000) Cholesterol binds to synaptophysin and is required for biogenesis of synaptic vesicles. Nat Cell Biol 2:42–49

Tsai HI, Tsai LH, Chen MY et al (2006) Cholesterol deficiency perturbs actin signaling and glutamate homeostasis in hippocampal astrocytes. Brain Res 1104:27–38

Wasser CR, Ertunc M, Liu X et al (2007) Cholesterol-dependent balance between evoked and spontaneous vesicle recycling. J Physiol 579(2):413–429

Xia F, Gao X, Kwan E et al (2004) Disruption of pancreatic β-cells lipid rafts modifies Kv2.1 channel gating and insulin exocyrtosis. J Biol Chem 279:24685–24691

Xia F, Leung YM, Gaisano G et al (2007) Targeting of Kv4, Cav1.2 and SNARE proteins to cholesterol-rich lipid rafts in pancreatic a-cells: effects on glucagons stimulus-secretion coupling. Endocrinology 148:2157–2167

Yancey PG, Rodrigueza WV, Kilsdonk EP et al (1996) Cellular cholesterol efflux mediated by cyclodextrins. J Biol Chem 271:16026–16034

Zamir O, Charlton MP (2006) Cholesterol and synaptic transmitter release at crayfish neuromuscular junctions. J Physiol 571:83–99

Zidovetzki R, Levitan I (2007) Use of cyclodextrins to manipulatre plasma membrane cholesterol content: evidence, misconceptions and control strategies. Biochim Biophys Acta 1768: 1311–1324

Chapter 3
Effects of Cholesterol-Depleting Agent Methyl-β-Cyclodextrin on the Functional State of Brain Nerve Terminals

Abstract Cholesterol acceptor methyl-β-cyclodextrin (MβCD) (15 mM) reduced the cholesterol content in brain nerve terminals (synaptosomes) by one quarter. The application of MβCD to the synaptosomes as well as isolated synaptic vesicles led to the gradual leakage of the protons from the vesicles, as shown by acridine orange fluorescence measurements, whereas the application of MβCD complexed with cholesterol (15:0.2) that increased the membrane cholesterol content stimulated additional vesicle acidification. It was supposed that cholesterol depletion of the plasma membrane with MβCD induced the removal of cholesterol from the membranes of synaptic vesicles resulting in immediate dissipation of synaptic vesicle proton gradient, and so could provoke redistribution of the neurotransmitter between the vesicular and cytosolic pools.

3.1 Extraction of Cholesterol from Brain Nerve Terminals by MβCD, Confocal Imaging and Flow Cytometric Analysis of Cholesterol-Deficient Nerve Terminals

Presynaptic nerve terminal promises to be one of the best systems to explore the relationship between the structure of a protein, its biochemical and cell-biological properties, and physiological role (Sudhof 2004). The synaptosomes, isolated brain nerve terminals, retain all features of intact nerve terminals, e.g., ability to maintain the membrane potential, exocytotic and transporter-mediated release as well as accomplish uptake of neurotransmitters. The synaptosomes were prepared by differential and Ficoll-400 density gradient centrifugation of rat brain homogenate (Fig. 3.1) according to the method of Cotman (1974) with slight modifications. Wistar rats (males 100–120 g body weight) were maintained in accordance with the European Guidelines and International Laws and Policies. The cerebral hemispheres of decapitated animals were rapidly removed and homogenized in ice-cold 0.32 M sucrose, 5 mM HEPES-NaOH (pH 7.4), and 0.2 mM EDTA. Then brain

T. Borisova, *Cholesterol and Presynaptic Glutamate Transport in the Brain*,   15
SpringerBriefs in Neuroscience 12, DOI 10.1007/978-1-4614-7759-4_3,
© Springer Science+Business Media New York 2013

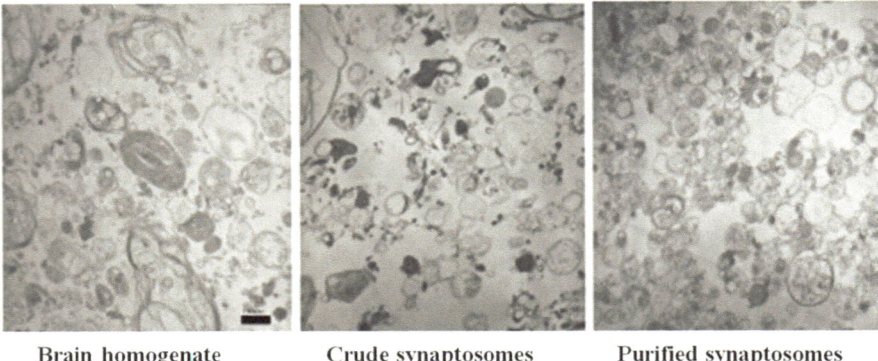

| Brain homogenate | Crude synaptosomes | Purified synaptosomes |

Fig. 3.1 Electron microscopy of brain homogenate, crude synaptosomes, and purified synaptosomes. Scale bar is 1 μm

Fig. 3.2 Confocal imaging of purified synaptosomes labeled with the fluorescent dye octadecylrhodamine B (R18). For confocal imaging, R18-labeled synaptosomes were evaluated using the confocal laser scanning microscope LSM 510 META, Carl Zeiss. Scale bar is 5 μm. Figure from Krisanova et al. (2009)

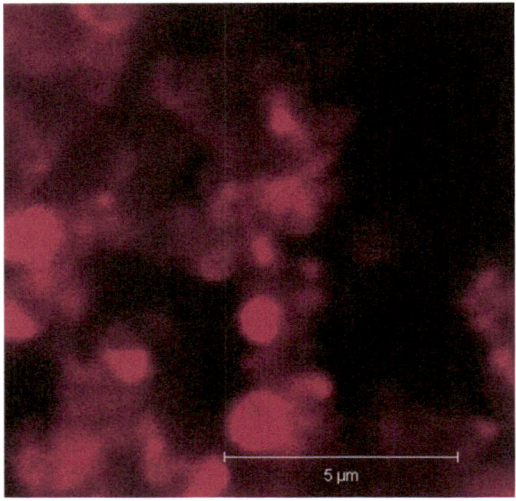

homogenate was subjected to differential and Ficoll-400 density gradient centrifugation and crude and purified synaptosomal fractions, respectively, were obtained (Figs. 3.1 and 3.2). The synaptosomal suspensions were used in experiments during 2–4 h after isolation. The standard salt solution was oxygenated and contained (in mM): NaCl 126; KCl 5; $MgCl_2$ 1.4; NaH_2PO_4 1.0; HEPES 20 (pH 7.4); and D-GLUCOSE 10. The Ca^{2+}-supplemented medium contained 2 mM $CaCl_2$. The Ca^{2+}-free medium contained 1 mM EGTA and no added Ca^{2+}. Protein concentration was measured as described by Larson et al. (1986).

Quantitative assessment of cholesterol concentration (see Appendix) showed that the treatment of the synaptosomes with 15 mM methyl-β-cyclodextrin (MβCD) for half an hour reduced the cholesterol level from 0.55 ± 0.02 μmol of cholesterol/mg of protein in the control to 0.42 ± 0.02 μmol of cholesterol/mg of protein after the treatment with the acceptor ($P \leq 0.05$, Student's t-test, $n = 4$); whereas cholesterol content of the synaptosomes treated with MβCD complexed with cholesterol ("neutral complex," 2.3 mM cholesterol in 15 mM MβCD) was not changed considerably, thereby making these synaptosomes appropriate for using as an additional control in the analysis of changes associated with cholesterol deficiency. The above data was confirmed using confocal laser scanning microscopy with the fluorescent probe filipin, which binds to membrane cholesterol (Kruth and Vaughan 1980). An immediate decrease in the fluorescence intensity of filipin just after application of MβCD under a confocal laser scanning microscope was revealed (Fig. 3.3a). The profiles of the fluorescence intensity of filipin represented in Fig. 3.3b showed that MβCD effectively extracted cholesterol from the synaptosomes (Krisanova et al. 2012).

Taking into account that intact synaptosomes could be identified and subjected to flow cytometric analysis without the loss of physiological responsiveness (Wolf and Kapatos 1989), flow cytometry was applied for comparative measurements of control and cholesterol-depleted synaptosomes to evaluate their size, cytoplasmic granularity, and the ability to maintain the membrane potential (Gylys et al. 2000; Mullaney et al. 1969). Flow cytometric studies revealed less than 10 % heterogeneity in the cell size (calculation based on MnX value) between the populations of control and cholesterol-depleted synaptosomes by forward scatter (FS) analysis that was a measure of light scattered at narrow angles. Cytoplasmic granularity measured by side scatter (SS) profiling (90° light scatter signal), which is a relative indicator of the fine internal structure (e.g., synaptic vesicles), also showed less than 10 % scatter of MnX between control and cholesterol-depleted synaptosomes (Fig. 3.4). Thus, flow cytometric analysis of control and cholesterol-depleted synaptosomes revealed similarity in their size and cytoplasmic granularity (Borisova et al. 2010a). It should also be noted that the populations of both control and cholesterol-depleted synaptosomes exhibited the ability to maintain the membrane potential that was demonstrated by significant peaks of the fluorescence of the potentiometric optical dye, rhodamine 6G.

The integrity of the synaptosomes was assessed in the presence of the cholesterol acceptor by monitoring the activity of the cytoplasmic enzyme lactate dehydrogenase (LDH) in the incubation medium. It was shown that LDH leakage from control synaptosomes and synaptosomes incubated with 15 mM MβCD for 40 min was almost similar. The activity of LDH in MβCD-treated synaptosomal suspension was higher by ~0.3 % as compared to the enzyme activity in the control, thereby indicating unchanged synaptosomal integrity after cholesterol depletion. These results are in a good agreement with the data of Wasser et al. (2007), who have not found significant breach of neuronal membrane integrity and cell viability after MβCD administration. Thus, no significant differences were revealed in the parameters tested, e.g., the size of the synaptosomes, cytoplasmic granularity, the plasma membrane potential, and extracellular LDH activity, between control and MβCD-treated synaptosomal preparations that allowed us to use cholesterol-depleted synaptosomes for further investigation of glutamate transport across the plasma membrane (Borisova et al. 2010a).

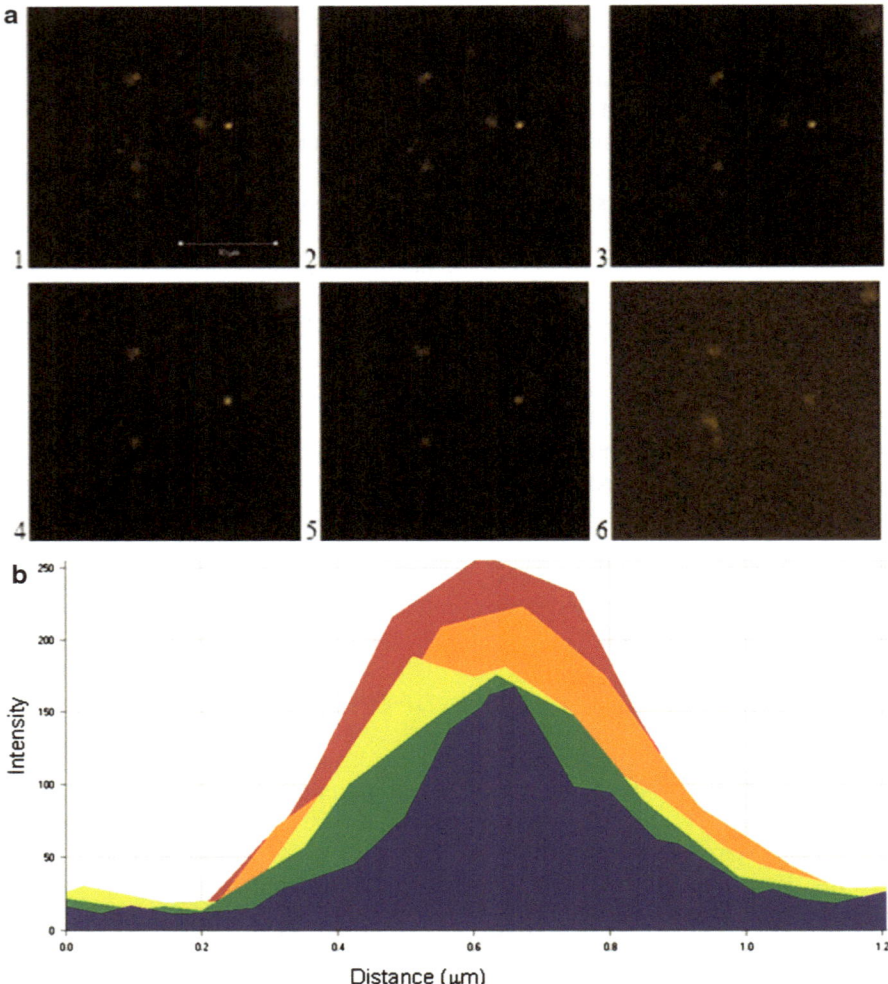

Fig. 3.3 (**a**) Confocal imaging of the synaptosomes labeled with filipin following the addition of methyl-β-cyclodextrin (MβCD). For confocal imaging, filipin-labeled synaptosomes were evaluated using the confocal laser scanning microscope LSM 510 META, Carl Zeiss (see Appendix). After starting the time series and the fluorescence images were captured with camera in each 4.5 s. Scale bar is 10 μm. Figure from Borisova et al. (2010a). (**b**) The profiles of the fluorescence intensity of cholesterol-sensitive fluorescent dye filipin recorded from the confocal images of filipin-labeled synaptosomes following the addition of MβCD. Figure from Krisanova et al. (2012)

The effect of MβCD on the planar lipid membrane (BLM) was analyzed. BLM was formed of solution of phosphatidylcholine and cholesterol (2:1) in n-heptane. After the addition of MβCD (10 mM) to both sides of BLM, current across the membrane in symmetrical solution consisted of 100 mM NaCl and 10 mM Tris–HCl pH 7.4 was not changed significantly (Fig. 3.5). This data is in a good agreement with above results on LDH.

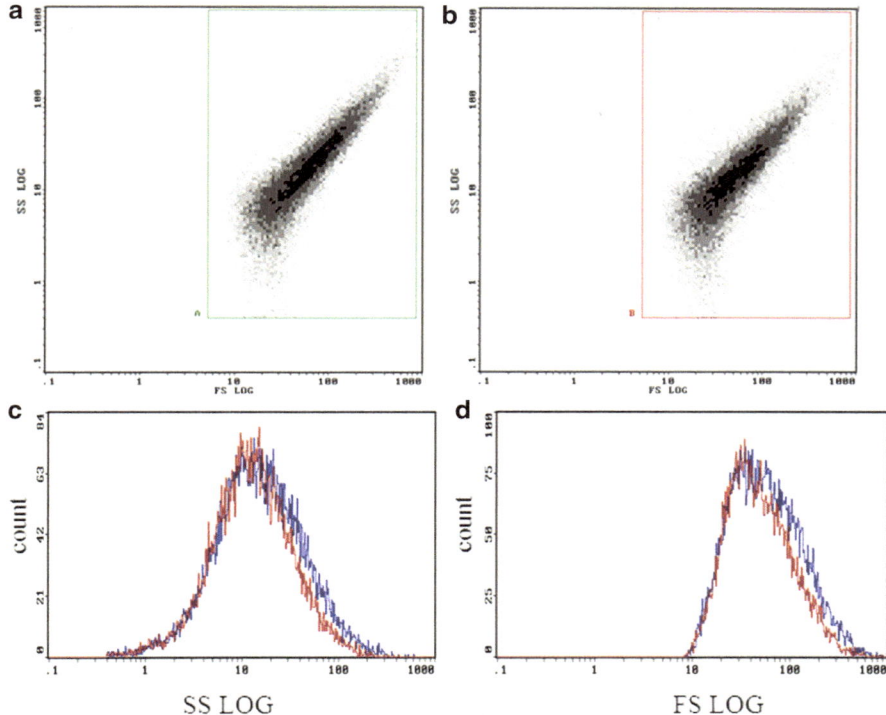

Fig. 3.4 Flow cytometric analysis of control ((**a**) *blue line* in plots (**c, d**)) and cholesterol-depleted synaptosomes ((**b**) *red line* in plots (**c, d**)). The measurements were performed on flow cytometer COULTER EPICS XL. (**a, b**) For each particle, FS and SS (90°) scatters are presented as *x* and *y* coordinates. (**c, d**) SS (**c**) and FS (**d**) scatters were represented as *x* coordinate. The analysis was based on 20,000 particles. All samples were examined with identical instrument settings. Figure from Borisova et al. (2010a)

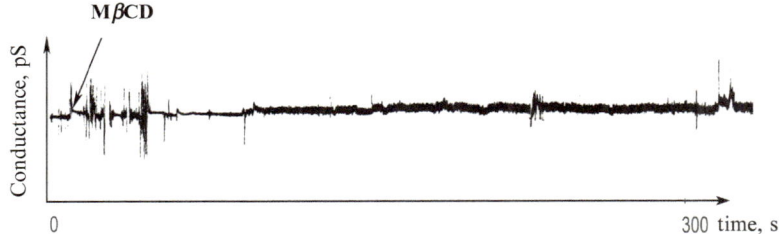

Fig. 3.5 The absence of the effect of MβCD on transmembrane current across planar lipid membrane (BLM) in symmetrical solution consisted of 100 mM NaCl and 10 mM Tris–HCl pH 7.4. Final concentration of MβCD consisted of 10 mM. Membrane potential was +100 mV (unpublished data of Dr. O. Shatursky and T. Borisova)

3.2   Effects of MβCD on the Proton Gradient of Synaptic Vesicles

Using a pH-sensitive fluorescent dye acridine orange, it was shown that the addition of 15 mM MβCD to the synaptosomes caused complete dissipation of the proton gradient across synaptic vesicles (Fig. 3.6, lines 1 and 2). The particular advantage of this methodological approach is a possibility to monitor not only vesicle acidification (i.e., ability of synaptic vesicles to keep protons) but also exo/endocytosis (Zoccarato et al. 1999). Also, no significant changes in either the intensity of the fluorescence signal or the emission spectrum of the dye were found on the addition of 5 mM MβCD to acridine orange solution and only a small jump in the fluorescence intensity (~6 %) in the presence of 15 mM MβCD (Tarasenko et al. 2010).

To visualize the processes occurring in synaptic vesicles during application of MβCD, synaptosomes were loaded with acridine orange and the changes in the fluorescent signal were looked at. Figure 3.7 demonstrates that the synaptosomes loaded with acridine orange are visualized as bright round spots clustered in groups. As can be seen from a time series of the images, the addition of MβCD (6 mM) to the chamber with the plated synaptosomes resulted in a decrease in the fluorescence intensity. Rapid and almost complete fading of the fluorescent signal, indicating the loss of acridine orange from vesicular lumen, was in a good agreement with spectrofluorometric measurements presented in Fig. 3.6 (Borisova et al. 2010b).

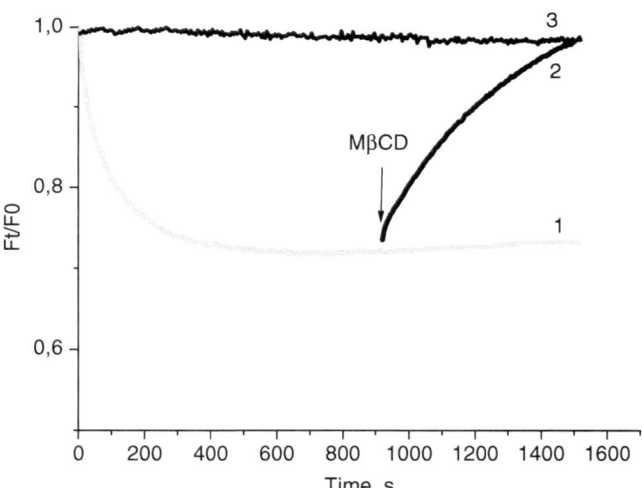

Fig. 3.6 The effects of MβCD on synaptic vesicle acidification. (1)—control synaptosomes. The synaptosomes (0.2 mg protein/mL) were incubated for 10 min at 30 °C and then equilibrated with acridine orange (5 µM). When the steady level of the dye fluorescence had been reached, 15 mM MβCD was added—(2); (3)—the synaptosomes in the presence of 15 mM MβCD, which was added 35 min before the addition of acridine orange. Each trace is representative of four experimental data records performed with different synaptosomal preparations. Figure from Borisova et al. (2010b)

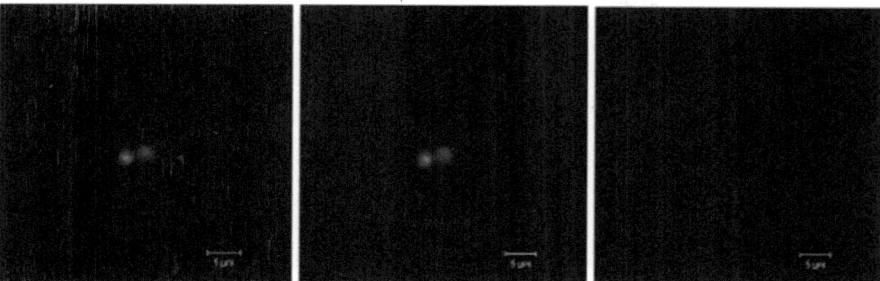

Fig. 3.7   Confocal imaging of the synaptosomes labeled with a pH-sensitive fluorescent dye acridine orange following the addition of MβCD. Measurements were performed by using the confocal laser scanning microscope LSM 510 META, Carl Zeiss. The synaptosomes were viewed in the absence of MβCD. Then MβCD (5 mM) was added to the thin layer of synaptosomal suspension squashed between glass surfaces and fluorescence images were captured with camera in each 5.77 s. In control experiments, no fading of the fluorescence signal of acridine orange-loaded synaptosomes was shown in this time interval. Scale bar is 5 μm. Figure from Borisova et al. (2010b)

If MβCD was incubated with the synaptosomes for 35 min and then removed from the synaptosomes by washing with 10 volumes of incubation media, the steady state level of the dye fluorescence, unexpectedly, was almost similar to the control (Fig. 3.6, similar to line 1). Thus, no significant alterations in synaptic vesicle acidification, i.e., the ability of the synaptosomes to accumulate acridine orange, were found after deletion of MβCD from the incubation media. The question rose whether the changes between acute and 30 min treatment were associated with the duration of the treatment and/or with the presence of the cholesterol acceptor in the incubating media. To assess this suggestion, synaptic vesicle acidification was analyzed during long-term extraction of cholesterol at 37 °C for 35 min, but 15 mM MβCD was not washed from the incubation media. Figure 3.6 (line 3) shows the inability of synaptic vesicles to accumulate acridine orange. Thus, the presence of MβCD in the incubation media prevented the restoration of the proton gradient of synaptic vesicles. The application of MβCD–cholesterol complex (15:2.3) caused insignificant changes in synaptic vesicle acidification at acute treatment, long-term pretreatment, and long-term treatment without washing of the acceptor (Borisova et al. 2010b).

The more probable cause of the alterations in the acidification of the synaptosomes is a reduction in proton pumping activity of V-type ATPase resulted from the changes in the cholesterol content of the synaptic vesicle membrane during MβCD-induced extraction of cholesterol from the plasma membrane of the synaptosomes. This assumption is supported by the data that a reduction of the cholesterol content inhibits the development of ΔpH by vacuolar H^+-ATPase (Perez-Castiñeira and Apps 1990) and by evidence that MβCD applied extracellularly can deplete cholesterol from the intracellular compartments (Lange et al. 1999; Zidovetzki and Levitan 2007). Whereas synaptic vesicles are cholesterol-rich organelles, it seems plausible that even small changes in the membrane cholesterol content may induce dramatic changes in the functional state of these structures. In this respect, it was interesting

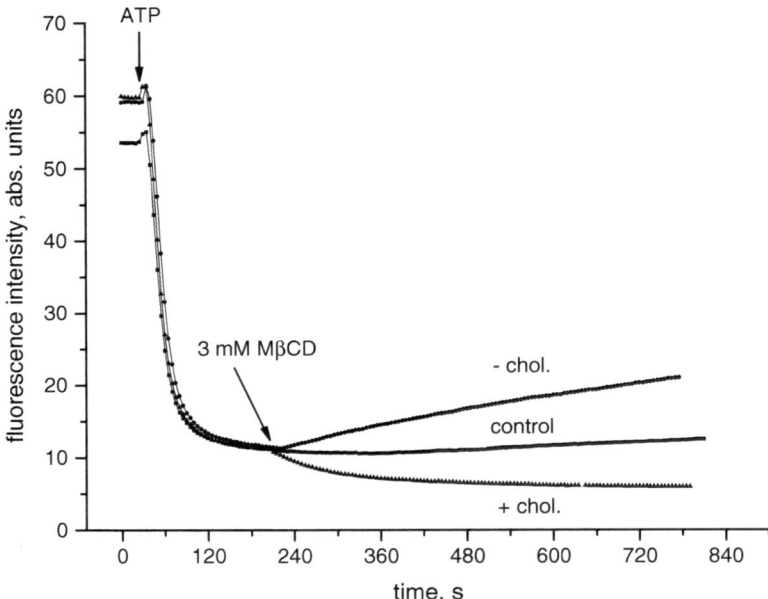

Fig. 3.8 The effect of MβCD, alone and complexed with cholesterol, on synaptic vesicle acidification. Three millimolar MβCD, alone or in complex with cholesterol (1:0.2), was added to isolated synaptic vesicles preincubated at 30 °C for 10 min in Ca^{2+}-containing medium and equilibrated with acridine orange. Traces are representative of three independent experiments. Figure from Tarasenko et al. (2010)

that the application of 3 mM MβCD in complex with cholesterol (1:0.2) to acridine orange-loaded synaptosomes not only prevented the leakage of acridine orange but also led to additional accumulation of the dye in synaptic vesicles, indicating an increase in proton pumping activity of vesicular H^+-ATPase. In this case, MβCD delivers cholesterol to the plasma membrane (Zamir and Charlton 2006), and thereafter extra cholesterol appears to be distributed between the intracellular compartments, including synaptic vesicles (Tarasenko et al. 2010).

High sensitivity of vesicle acidification to the level of cholesterol in the vesicle membrane was confirmed in the experiments with isolated synaptic vesicles. As seen in Fig. 3.8, MβCD (3 mM) itself caused the leakage of the dye (and so protons) from acridine orange-loaded synaptic vesicles, whereas MβCD complexed with cholesterol induced additional acidification of the vesicles (Tarasenko et al. 2010).

Dissipation of the proton gradient of the acidic compartments of the cells evoked by MβCD was also shown in non-nervous cells. Using pH-sensitive fluorescent dye acridine orange, it was demonstrated that the acidification of secretory granules of human and rabbit platelets was decreased by ~15 % and 51 % after the addition of 5 and 15 mM MβCD, respectively; whereas the enrichment of platelet plasma membrane with cholesterol by the complex MβCD–cholesterol (1:0.2) led to the additional accumulation of acridine orange in secretory granules indicating an increase in the proton pumping activity of vesicular H^+-ATPase (Borisova et al. 2011).

References

Borisova T, Krisanova N, Sivko R, Borysov A (2010a) Cholesterol depletion attenuates tonic release but increases the ambient level of glutamate in rat brain synaptosomes. Neurochem Int 56:466–478

Borisova T, Sivko R, Borysov A, Krisanova N (2010b) Diverse presynaptic mechanisms underlying methyl-beta-cyclodextrin—mediated changes in glutamate transport. Cell Mol Neurobiol 30:1013–1023

Borisova T, Kasatkina L, Ostapchenko L (2011) The proton gradient of secretory granules and glutamate transport in blood platelets during cholesterol depletion of the plasma membrane by methyl-beta-cyclodextrin. Neurochem Int 59:965–975

Cotman CW (1974) Isolation of synaptosomal and synaptic plasma membrane fractions. Methods Enzymol 31:445–452

Gylys K, Fein J, Cole GM (2000) Quantitative characterization of crude synaptosomal fraction (P-2) components by flow cytometry. J Neurosci Res 61:186–192

Krisanova NV, Ttiksh IO, Borisova TA (2009) Synaptopathy under conditions of altered gravity: changes in synaptic vesicle fusion and glutamate release. Neurochem Int 55:724–731

Krisanova N, Sivko R, Kasatkina L, Borisova T (2012) Neuroprotection by lowering cholesterol: a decrease in membrane cholesterol content reduces transporter-mediated glutamate release from brain nerve terminals. Biochim Biophys Acta 1822:1013–1023

Kruth HS, Vaughan M (1980) Quantification of low density lipoprotein binding and cholesterol accumulation by single human fibroblasts using fluorescence microscopy. J Lipid Res 21:123–130

Lange Y, Ye J, Rigney M, Steck TL (1999) Regulation of endoplasmic reticulum cholesterol by plasma membrane cholesterol. J Lipid Res 40:2264–2269

Larson E, Howlett B, Jagendorf A (1986) Artificial reductant enhancement of the Lowry method for protein determination. Anal Biochem 155:243–248

Mullaney PF, Van Dilla MA, Coulter JR (1969) Cell sizing: a light scattering photometer for rapid volume determination. Rev Sci Instrum 40:1029–1032

Perez-Castiñeira JR, Apps DK (1990) Vacuolar H(+)-ATPase of adrenal secretory granules. Rapid partial purification and reconstitution into proteoliposomes. Biochem J 271:127–131

Sudhof TC (2004) The synaptic vesicle cycle. Annu Rev Neurosci 27:509–547

Tarasenko AS, Sivko RV, Krisanova NV, Himmelreich NH, Borisova TA (2010) Cholesterol depletion from the plasma membrane impairs proton and glutamate storage in synaptic vesicles of nerve terminals. J Mol Neurosci 41:358–367

Wasser CR, Ertunc M, Liu X et al (2007) Cholesterol-dependent balance between evoked and spontaneous vesicle recycling. J Physiol 579(2):413–429

Wolf M, Kapatos G (1989) Flow cytometric analysis of rat striatal nerve terminals. J Neurosci 9:94–105

Zamir O, Charlton MP (2006) Cholesterol and synaptic transmitter release at crayfish neuromuscular junctions. J Physiol 571:83–99

Zidovetzki R, Levitan I (2007) Use of cyclodextrins to manipulatre plasma membrane cholesterol content: evidence, misconceptions and control strategies. Biochim Biophys Acta 1768:1311–1324

Zoccarato F, Cavallini L, Alexandre A (1999) The pH-sensitive dye acridine orange as a tool to monitor exocytosis/endocytosis in synaptosomes. J Neurochem 72:625–633

Chapter 4
The Extracellular Level and Uptake of Glutamate in Cholesterol-Deficient Nerve Terminals

Abstract The low level of ambient glutamate is extremely important for the brain's spontaneous activity and proper synaptic transmission. Cholesterol deficiency has been implicated in the pathogenesis of several neurodegenerative disorders. It was examined whether membrane cholesterol modulated the extracellular glutamate level in the nerve terminals and the processes responsible for its maintenance. The ambient L-[^{14}C]glutamate level, being an equilibrium between Na$^+$-dependent uptake and tonic release, was increased from 0.193 ± 0.013 nmol/mg protein to 0.282 ± 0.013 nmol/mg protein (extracellular endogenous glutamate—from 6.9 ± 2.0 nmol/mg protein to 16.6 ± 2.0 nmol/mg protein, respectively) in rat brain synaptosomes treated with a cholesterol acceptor MβCD. This alteration was not due to the change in the activity of glutamine synthetase that was shown with the specific blocker L-methionine sulfoximine. Also, cholesterol-deficient synaptosomes showed a lower initial velocity of L-[^{14}C]glutamate uptake.

The most commonly used methodological approach for the evaluation of the extracellular glutamate level is the measurements in slices and in vivo microdialysis (Jabaudon et al. 1999; Baker et al. 2002; Cavelier and Attwell 2005; Nyitrai et al. 2006; Meur et al. 2007). However, in the contex of diverse effects of cholesterol depletion on neurons and glial cells (Butchbach et al. 2004; Tsai et al. 2006), these measurements do not clarify the contribution of each type of the cells in the maintenance of the extracellular level of glutamate in cholesterol deficiency.

As it was mentioned above, some studies on analysis of the role of cholesterol in synaptic transmission were carried out in the presence of MβCD in the incubation medium, the other ones—after its removal from the medium. In addition, in different methods of MβCD application, the incubation time was also different. Also, from the previous chapter it is clear that the presence of the acceptor in the incubation media is not indifferent to the synaptosomes. So, it may be hypothesized that the mechanisms underlying the changes in key processes of synaptic transmission under conditions of cholesterol deficiency may be diverse in different

T. Borisova, *Cholesterol and Presynaptic Glutamate Transport in the Brain*, SpringerBriefs in Neuroscience 12, DOI 10.1007/978-1-4614-7759-4_4, © Springer Science+Business Media New York 2013

Fig. 4.1 The acute treatment of the synaptosomes in the presence of MβCD (AT) and long-term pretreatment of the synaptosomes with the acceptor followed by its washing from the incubation medium (LP). C-control synaptosomes

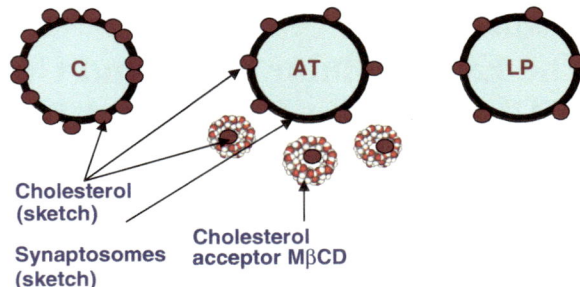

Cholesterol (sketch)

Synaptosomes (sketch)

Cholesterol acceptor MβCD

methodological protocols of the application of the acceptor. In these studies, three main methodological approaches were used for treatment of the synaptosomes with MβCD for extraction of membrane cholesterol. The first methodical approach—the acute treatment of the synaptosomes in the presence of the acceptor (AT), when the parameters were analyzed immediately after the addition of MβCD to the incubation medium (Fig. 4.1). To determine whether the difference in MβCD actions related to different incubation time or the presence of acceptor in the incubation medium, the second protocol of incubation was used. In some experiments, the more detailed analysis of MβCD influence on L-[^{14}C]glutamate transport was performed using the long-term (35 min) treatment of the synaptosomes with MβCD without washing of the acceptor (LT). The third methodological approach is long-term (35 min) pretreatment of the synaptosomes with the acceptor followed by its washing from the incubation medium by a tenfold dilution and centrifugation of the synaptosomes (LP) (Fig. 4.1).

4.1   Effects of Acute and Long-Term Treatment of the Synaptosomes with MβCD on Glutamate Uptake

It was shown that 15 mM MβCD acutely applied to the synaptosomes decreased the initial velocity of L-[^{14}C]glutamate (10 μM) uptake by 64 % that was equal to 2.5 ± 0.3 nmol$\times$min$^{-1}\times$mg^{-1} of proteins in the control and 0.9 ± 0.1 nmol$\times$min$^{-1}\times$mg^{-1} in the presence of the acceptor ($P \leq 0.05$, $n=6$). LT with 15 mM MβCD also caused a decrease in the initial rate of uptake that was equal to 2.5 ± 0.3 nmol$\times$min$^{-1}\times$mg^{-1} of proteins and 1.5 ± 0.2 nmol$\times$min$^{-1}\times$mg^{-1} of proteins, respectively ($P \leq 0.05$, $n=8$). The synaptosomes were also treated with 15 mM MβCD complexed with cholesterol (2.3 mM) to determine whether the observed MβCD-induced effects were a result of cholesterol extraction. Application of MβCD-cholesterol complex caused insignificant changes in L-[^{14}C]glutamate uptake by synaptosomes during AT and LT as compared to the control, thereby showing that decreased uptake registered after MβCD treatment was associated with depletion of membrane cholesterol (Borisova et al. 2010b).

Synaptosomal glutamate uptake was also analyzed during long-term extraction of cholesterol at 37 °C for 35 min, but MβCD was not washed from the incubation media. It was revealed that 15 mM MβCD decreased the initial velocity of L-[^{14}C]

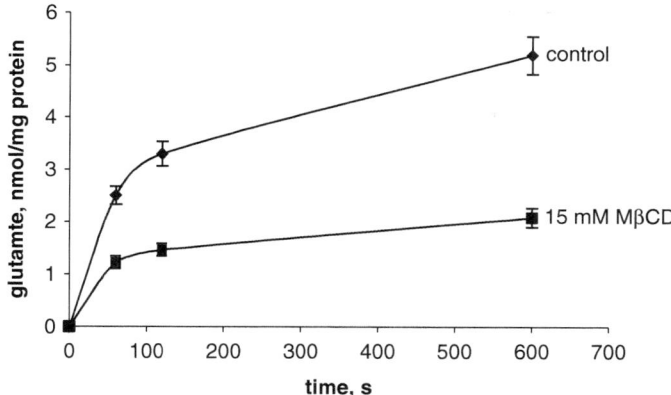

Fig. 4.2 The reduction of glutamate uptake during cholesterol removal. The time course of uptake of L-[^{14}C]glutamate during application of 15 mM MβCD for 35 min at 37 °C without consequent washing of the acceptor. Uptake was initiated by the addition of 10 μM L-glutamate supplemented with 420 nM L-[^{14}C]glutamate (0.1 μCi/mL). The samples were incubated at 37 °C for 1, 2, and 10 min, and then rapidly centrifuged. L-[^{14}C]glutamate radioactivity in the aliquots of the supernatant and pellet were determined. Data represent the mean±SEM of five independent experiments performed in triplicate. Figure from Borisova et al. (2010b)

glutamate (10 μM) uptake by more than 60 % under this conditions (Fig. 4.2). These data allowed suggesting that the changes in glutamate uptake were associated not only with the direct influence of cholesterol depletion on glutamate transporters (as it was after LT), but also with dissipation of the proton gradient shown in the previous chapter, Fig. 3.6, line 3. It should be noted that accumulation of L-[^{14}C]glutamate (10 μM) for 10 min was also changed in a similar manner with the initial velocity of L-[^{14}C]glutamate uptake at different protocols of MβCD application (Borisova et al. 2010b).

4.2   The Extracellular Glutamate Level After Cholesterol Depletion

It is clear that a low level of ambient glutamate is extremely important for the brain's spontaneous activity and proper synaptic transmission and even small changes in the ambient level of the neurotransmitter may significantly influence CNS functions (Cavelier and Attwell 2005; Westphalen and Hemmings 2006). Ambient L-[^{14}C]glutamate was drastically increased just after the addition of MβCD to the synaptosomes (AT) and became 2 times higher at 10 min time point in the presence of 15 mM MβCD as compared to the control (Fig. 4.3). This increase was underscored by dramatic leakage of L-[^{14}C]glutamate from the nerve terminals in response to the acceptor application that was equal to 0.018 ± 0.001 nmol×mg protein^{-1}×min^{-1} (for 15 mM MβCD), whereas in the control this value consisted of 0.001 ± 0.0005 nmol×mg protein^{-1}×min^{-1} ($P \leq 0.05$, Student's t-test, $n=4$) (Borisova et al. 2010b).

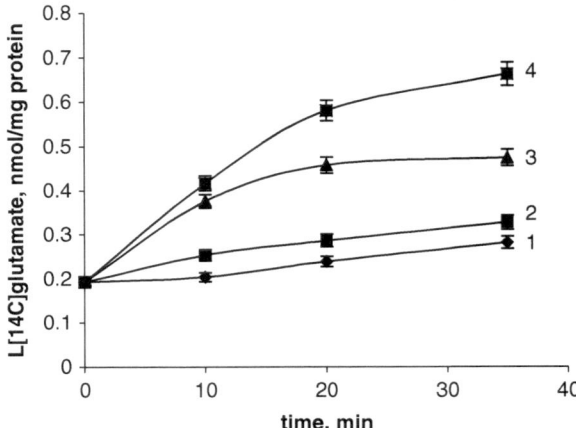

Fig. 4.3 The extracellular level of L-[^{14}C]glutamate in the synaptosomes after the addition of 0, 5, 15, and 30 mM MβCD—curves 1; 2; 3; and 4, respectively. The synaptosomes were suspended in oxygenated ice-cold standard salt solution to final protein concentration of 0.2 mg/mL, incubated for 10 min at 37 °C, and then MβCD were applied in different concentrations. The samples were centrifuged at different time points and L-[^{14}C] glutamate radioactivity in the supernatants was determined. The amount of radiocarbon released into the medium was expressed as percentage of total radioactivity in the synaptosomes. Total synaptosomal L-[^{14}C] glutamate content was equal to 200,000 ± 15,000 cpm/mg protein. Data are means ± SEM of five independent experiments, each performed in triplicate. Figure from Borisova et al. (2010b)

In LT, the extracellular level of L-[^{14}C]glutamate became one third higher after the treatment and consisted of 0.193 ± 0.013 nmol/mg protein in the control and 0.282 ± 0.013 nmol/mg protein in 15 mM MβCD-treated synaptosomes ($P \leq 0.05$, Student's t-test, $n = 8$) (Fig. 4.4, the first and second columns). Continuous monitoring showed that the extracellular level of L-[^{14}C]glutamate remained enhanced in cholesterol-depleted synaptosomes at least during 30 min time period. Whereas, the synaptosomes treated with 15 mM MβCD complexed with cholesterol (2.3 mM) exhibited only insignificant changes in extracellular L-[^{14}C]glutamate as compared to untreated control (Fig. 4.4, the third column). Thus, it would be expected that an increase in the ambient glutamate level in MβCD-treated synaptosomes was a result of depletion of membrane cholesterol, but not the effect of MβCD per se irrespective to cholesterol accepting capacity (Borisova et al. 2010a).

For the confirmation of the accuracy of the measurements with preloaded L-[^{14}C] glutamate, ambient endogenous glutamate was assessed in the control and cholesterol-depleted synaptosomes using Amino Acid Analyzer. A much more significant difference was revealed in extracellular endogenous glutamate between control and cholesterol-depleted synaptosomes, than it was measured with preloaded L-[^{14}C]glutamate. The extracellular content of endogenous glutamate was more than 2 times lower in control synaptosomes in comparison with MβCD-treated ones and consisted of 6.9 ± 2.0 nmol/mg protein in control and 16.6 ± 2.0 nmol/mg protein after

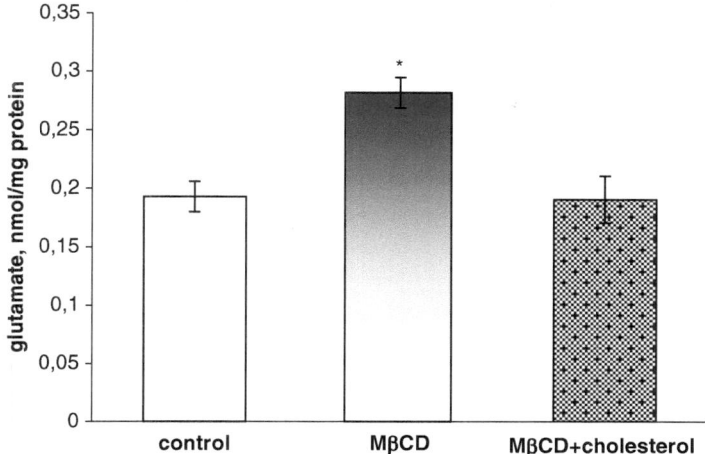

Fig. 4.4 The extracellular L-[14C]glutamate level of control (*empty bar*), cholesterol-depleted synaptosomes (*shaded bar*) and the synaptosomes treated with MβCD complexed with cholesterol (*grey dotted bar*). The synaptosomes were incubated without MβCD (the control experiments), with 15 mM MβCD (extraction of cholesterol) or with 15 mM MβCD complexed with 2.3 mM cholesterol (analysis of the effects of MβCD irrespective to cholesterol accepting capacity) at 37 °C during 30 min followed by washing. Control, MβCD- and MβCD/cholesterol-treated synaptosomes were loaded with L-[14C]glutamic acid (500 nM, 238 mCi/mmol) in Ca^{2+}-supplemented oxygenated standard salt solution. After loading, the extracellular level of L-[14C]glutamate was measured according to the following method: samples (125 μL of the suspension, 0.5 mg of protein/mL) were preincubated for 8 min at 37 °C to restore ion gradients (the common procedure for all measurements), then the preparations were incubated further for 6 min at 37 °C and rapidly sedimented in a microcentrifuge. L-[14C] glutamate radioactivity in the supernatants was determined. Total synaptosomal L-[14C] glutamate content was equal to 200,000 ± 15,000 cpm/mg protein. Data are means ± SEM of eight independent experiments, each performed in triplicate. Data are compared by Student's *t*-test. *$P \le 0.05$ as compared to the control or the synaptosomes treated with MβCD complexed with cholesterol. Figure from Borisova et al. (2010a)

cholesterol depletion ($P \le 0.05$, Student's *t*-test, $n = 4$) (Fig. 4.5, the first couple of the columns). Thus, the analysis of the ambient level setting by endogenous glutamate revealed that it became higher after cholesterol depletion, thereby confirming our data with preloaded L-[14C]glutamate (Borisova et al. 2010a).

Glutamate dehydrogenase assay was also used for the assessment of the extracellular level of endogenous glutamate in cholesterol-deficient nerve terminals. As shown in Fig. 4.6, the ambient level of endogenous glutamate was ~35 % higher in cholesterol-deficient nerve terminals in comparison with the control ones. The application of 5 mM MβCD complexed with cholesterol (2.3 mM) caused insignificant changes in the extracellular glutamate level in the synaptosomes as compared to the control (Krisanova et al. 2012). Thus, the results on the assessment of the extracellular level of glutamate in cholesterol-deficient synaptosomes obtained with glutamate dehydrogenase assay were in accordance with our data on radiolabeled and endogenous glutamate.

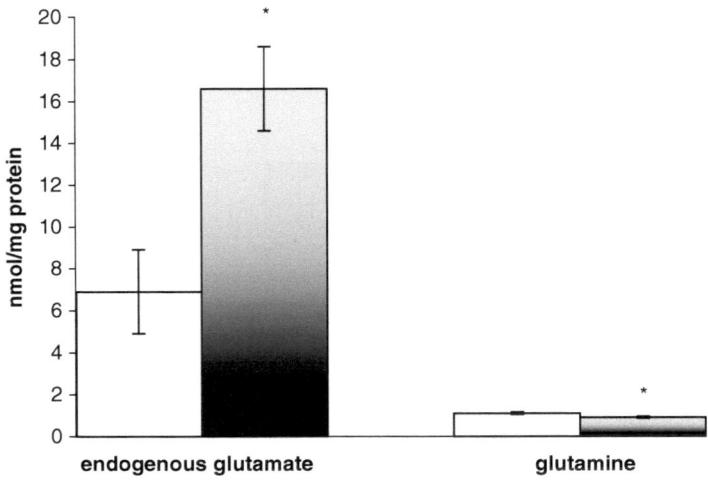

Fig. 4.5 The extracellular level of endogenous glutamate and glutamine in control (*empty bar*) and cholesterol-depleted synaptosomes (*shaded bar*). For extraction of cholesterol, the synaptosomes (15 mL of the suspension, 2 mg of protein/mL) were incubated with 15 mM MβCD at 37 °C during 30 min followed by washing. For the evaluation of the ambient level of the neuromediator, the synaptosomes (5 mL of the suspension, 6 mg of protein/mL) controlled or treated with MβCD were incubated at 37 °C for 15 min, and then sedimented. The supernatants were concentrated in rotor evaporator more than ~2.5 times. Extracellular endogenous glutamate and glutamine in the synaptosomes were measured using Amino Acid Analyzer. Data are means ± SEM of four independent experiments, each performed in triplicate. Data are compared in each duplex separately by Student's *t*-test. *$P \leq 0.05$ as compared to each control. Figure from Borisova et al. (2010a)

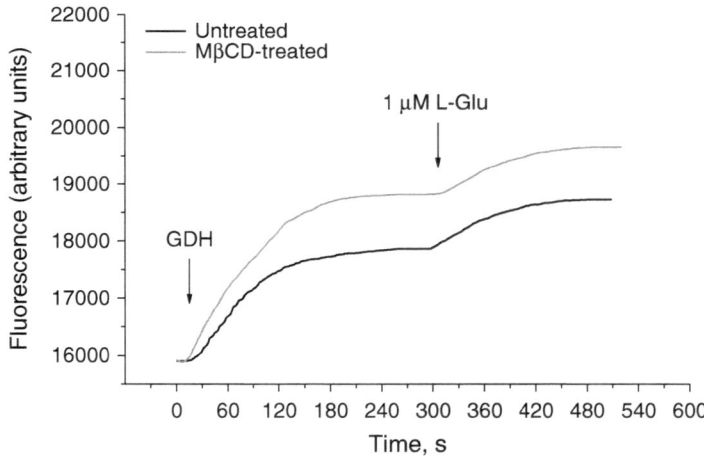

Fig. 4.6 The extracellular level of endogenous glutamate in control (*black line*) and cholesterol-deficient (*grey line*) rat brain synaptosomes assessed with glutamate dehydrogenase assay. Control and 15 mM MβCD-treated synaptosomal suspension (0.5 mg/mL of final protein concentration) was added to an enzymatic assay solution containing glutamate dehydrogenase (GDH). The extracellular level of endogenous glutamate in the synaptosomes was measured by the changes in NADH fluorescence (excitation and emission wavelengths of 340 and 460 nm, respectively). Trace is representative of three independent experiments. Figure from Krisanova et al. (2012)

The reaction that transforms glutamate in glutamine is ensured by the cytoplasmic enzyme glutamine synthetase, which specific activity is several times lower in neurons in comparison with that observed in astrocytes (Patel et al. 1983; Tansey et al. 1991). Applying the glutamine synthetase blocker L-methionine sulfoximine (MSO, 1.5 mM) to the synaptosomes preliminary loaded with L-[^{14}C]glutamate, it was revealed that in the presence of the inhibitor, the extracellular glutamate level was higher by ~30 % in cholesterol deficiency as compared to the control. Moreover, the treatment with MSO extended the difference in ambient glutamate between control and cholesterol-depleted synaptosomes. Thus, the enhanced level of ambient L-[^{14}C] glutamate in cholesterol deficiency was not originated from the alteration in the activity of glutamine synthetase. Also, namely L-[^{14}C]glutamate (but not L-[^{14}C] glutamine) determined the augmented extracellular level of radioactivity in MβCD-treated synaptosomes (see Fig. 4.4) that was in accordance with the above analysis of the ambient level of endogenous glutamate shown in the first couple of the columns of Fig. 4.5. Studying MSO effects with the other experimental protocol when 1.5 mM MSO was applied before L-[^{14}C]glutamate loading procedure, it was also revealed that the extracelular level of the neurotransmitter was higher after cholesterol depletion than in the control. Taking into account that the treatment with MSO before and after L-[^{14}C]glutamate loading did not mask the difference in the level of ambient glutamate between control and cholesterol-depleted synaptosomes, it was suggested that the inhibition of enzyme activity was not the main cause of enhanced ambient glutamate in cholesterol deficiency. Also, it seems interesting that the administration of MSO to control synaptosomes before L-[^{14}C]glutamate loading caused a decrease in the extracellular glutamate level by 30 % as compared to the synaptosomes without the inhibitor. Whereas in MβCD-treated synaptosomes, MSO did not attenuate the ambient glutamate content. It may be supposed from these series of the experiments that glutamine synthetase per se is sensitive to depletion of membrane cholesterol. Our data on the assessment of the ambient level of endogenous glutamine by Amino Acid Analyzer may be considered in support of this suggestion. The extracellular level of endogenous glutamine was lesser by ~20 % in cholesterol-deficient synaptosomes as compared to the control and consisted of 1.1 ± 0.05 nmol/mg protein in the control and 0.9 ± 0.05 nmol/mg protein after the treatment ($P \leq 0.05$, Student's t-test, $n=4$) (Fig. 4.5, the second couple of the columns). It was assumed that the latest changes could result from the inhibition of the activity of glutamine synthetase after cholesterol depletion (Borisova et al. 2010a).

The analysis of ambient endogenous glutamate/glutamine and the experiments with glutamine synthetase blocker MSO confirmed an increase in the extracellular level of glutamate in cholesterol-deficient synaptosomes. It should be noted that the ambient level of glutamine measured by the Amino Acid Analyser was found to be lesser after cholesterol depletion, thereby indicating that the latest might also influence glutamate/glutamine cycle. These data were in accordance with the results of Tsai et al. (2006), where the activity of glutamine synthetase was shown to be suppressed in cultured astrocytes after cholesterol reduction. Using MSO, it was clearly demonstrated that the inhibition of glutamine synthetase activity did not eliminate the difference in ambient glutamate between control and cholesterol-deficient synaptosomes, and thus changing enzyme activity was not the main cause of an increase in extracellular glutamate in cholesterol deficiency (Borisova et al. 2010a).

The next experiments were focused on other possible mechanisms underlying the enhancement of the extracellular glutamate level in cholesterol deficiency. The average ambient level of glutamate is set by a balance between the rate of transporter-mediated uptake and the rate of tonic release of glutamate that can be represented as follows:

$$[\text{extracellular glutamate}] = [\text{tonically released glutamate}] - [\text{glutamate uptaken by transporters}].$$

The next question asked was whether an increase in the extracellular glutamate level was a result of the changes in uptake and/or tonic release of glutamate after cholesterol depletion. The last term on the right side of the equation is ensured by Na^+-dependent glutamate transporters, which remove glutamate from the extracellular space. As it was shown above, cholesterol extraction caused a significant decrease (41 %) in the initial velocity of synaptosomal uptake of 10 µM L-[^{14}C] glutamate. Thus, significant lowering of the initial velocity of glutamate uptake but not the changes in glutamine synthetase activity was at least one of the causes of the elevation of the extracellular glutamate level in cholesterol-depleted synaptosomes. However, the equation in the middle of this section indicated that glutamate uptake activity was not the only factor, which determined the extracellular glutamate level. Tonic release of glutamate can also significantly contribute to an increase in the extracellular glutamate level, thereby the further attention was focused on the assessment of the tonic release value in cholesterol-deficient synaptosomes.

It was shown in this chapter that AT and LT treatment with 15 mM MβCD caused a significant decrease in the initial velocity of glutamate uptake and accumulation of glutamate by the synaptosomes for 10 min. However, only after LT a decrease in glutamate uptake was caused by the changes in functioning of glutamate transporters per se, possibly due to the alterations in their organization and trafficking in the plasma membrane. The presence of the acceptor in the incubation media was associated with dissipation of the proton gradient (Fig. 3.6 lines 2 and 3). AT and the long-term treatment without washing of MβCD were accompanied with the loss of synaptic vesicle acidification, thereby decreasing Na^+-dependent glutamate uptake (Fig. 4.2) possibly due to lower accumulation of glutamate by synaptic vesicles. It should be underlined that dissipation of the proton gradient, which is a driving force for glutamate accumulation by synaptic vesicles, caused a decrease in glutamate uptake by Na^+-dependent transporters of the plasma membrane that was shown with bafilomycin, a highly specific inhibitor of V-type ATPase (Borisova et al. 2010b).

Taking into consideration the great importance of the low level of extracellular glutamate for proper synaptic transmission, special attention was paid to the assessment of the ambient glutamate level in cholesterol deficiency. The baseline of the extracellular glutamate concentration in the CNS could have a major effect on neuronal excitability and synaptic transmission and even small changes in the ambient level of the neurotransmitter may significantly influence CNS functions (Cavelier and Attwell 2005; Westphalen and Hemmings 2006). AT with 15 mM MβCD was accompanied with a drastic increase in the extracellular glutamate level (Fig. 4.3). It became twice as much higher as compared to the control, whereas after LT with 15 mM MβCD the ambient glutamate concentration was increased not so considerably (by one third) as compared to the control. An augmentation of [glutamate]$_{out}$ during AT resulted from a significant increase in tonic (unstimulated) release and a decrease in uptake of the neurotransmitter (In contrast to AT, it was considered that the augmented ambient glutamate level after LT was set by the malfunction of glutamate transporters only, because tonic release became lower (see below).).

The finding that the significant level of cholesterol in the nerve terminals is required for maintaining a low extracellular glutamate concentration and the latter can be modulated by the changes in membrane cholesterol is an important observation emerging from this study. The proper concentration of ambient neurotransmitter is functionally important for tonic activation of excitatory and inhibitory post- and presynaptic receptors, so the increased level of extracellular glutamate in cholesterol-deficient synaptosomes can have profound consequences for synaptic transmission. Sah et al. (1989) showed that NMDA receptors in pyramidal cells of hippocampal slices could be tonically activated by the background level of glutamate present in the extracellular space. Dentate gyrus granule cell NMDA receptors were activated by ambient glutamate and generated an inward current, which increased the excitability of the neurons and was suppressible by NMDA receptor blockers (Dalby and Mody 2003). The effects of prolonged inhibition of glutamate uptake by the glutamate transport blocker, L-*trans*-pyrrolidine-2,4-dicarboxylate, on the relative abundance of mRNAs coding for NMDA receptor subunits, and the expression of corresponding proteins were investigated in the primary cultures of rat cerebellar granule neurons. It was shown that expression of NR2A and NR2B mRNAs was 40–50 % lower in the L-*trans*-pyrrolidine-2,4-dicarboxylate-treated cells as compared to the control. NR2B subunits are of importance for NMDA receptor localization and endocytosis, and it is suggested that NR2B-containing NMDA receptors play a role in the underlying pathophysiology of the neurodegenerative disorders, such as Alzheimer's and Huntington's diseases. It was suggested that reduced glutamate uptake resulting in the increased concentration of ambient glutamate initiated a series of adaptive responses manifested as a gradual downregulation of the functional activity and expression of NMDA receptors (Cebers et al. 2001). Since NMDA receptors are only partially involved in basal synaptic transmission, the importance of kainite and AMPA receptors are highlighted by the fact that the impairment of kainite receptor has a pivotal role in a decrease in basal synaptic transmission (Frank et al. 2008). Activation of presynaptic mGluRs in hippocampus was found to inhibit glutamate release (Forsythe and Clements 1990) or to facilitate it (McBain et al. 1994) at glutamate concentrations less than 1 μM that was equal to the ambient glutamate concentrations found in vivo. However, the increased level of ambient glutamate induced desensitization of NMDA, AMPA, and kainite receptors, thereby confirming that an alteration in the ambient glutamate concentration could have profound consequence for the information processing carried out by neurons (Cavelier et al. 2005). There is increasing evidence that the spontaneous activity of the brain provides a context for the analysis of incoming sensory signals (Cavelier et al. 2005; Kenet et al. 2003). In patients with medically intractable mesial temporal lobe epilepsy, an excess of extracellular glutamate in the hippocampus has been linked to the generation of recurrent seizures. Also, the augmented level of extracellular glutamate in the brain induced by cerebral ischemia led to neuronal death, mainly through overactivation of NMDA receptors (Ponce et al. 2008).

As it was already mentioned, astroglial glutamate transporters contributed significantly to the establishment of the definite concentration of extracellular glutamate in vivo (Jabaudon et al. 1999; Cavelier and Attwell 2005). In contrast to the neurons, cholesterol depletion increased the activity of glutamate transporters in astrocytes (Butchbach et al. 2004; Tsai et al. 2006). Thus, they could eliminate the changes in ambient glutamate associated with the malfunction of neuronal glutamate transporters in cholesterol deficiency that might be considered as a compensatory mechanism underlying synaptic plasticity. However, despite the augmented activity of astroglial glutamate transporters, and their increased ability to accumulate the neuromediator in cholesterol deficiency, the reduction in the glutamine synthetase activity might increase leakage of glutamate from astrocytes. This leakage could contribute to the enlargement of the extracellular glutamate concentration after cholesterol depletion, thereby further increasing ambient glutamate originated from the neurons. However, against this suggestion was the fact that the inhibition of glutamine synthetase in astrocytes caused a decrease in the glutamate concentration of the extracellular space (Ohnishi et al. 1995).

Taking into consideration the above experimental data, it was suggested that an increase in the ambient glutamate concentration impairing synaptic transmission could form the potential basis for the neurological symptoms of the diseases associated with the alterations in cholesterol homeostasis. A number of studies reported a reduced level of cellular cholesterol in Alzheimer's disease brain and in the mouse model for Alzheimer's disease, partly due to oligomeric amyloid β-protein causing lipid leakage from the neural cells (Mason 1994; Svennerholm and Gottfries 1994; Yao and Papadopoulos 2002; Roher et al. 2002; Michikawa 2003; Molander-Melin et al. 2005). Several data indicated that the brain cells deficient in cholesterol were less capable of withstanding neurological insults (Nicoll et al. 1995; Teasdale et al. 1997). Naturally occurring disorder Niemann-Pick type C1 was characterized by the lower concentration of cholesterol in neurons due to a decrease in the efficiency of cholesterol trafficking (Karten et al. 2002, 2003). Patients with this disorder demonstrated the neurological symptoms and neurodegeneration. We suggested that the latest might be associated with an elevation of the extracellular glutamate level. It should be also kept in mind that the neurological symptoms in other diseases could be provoked by the changes in the cholesterol level indirectly involved in the pathogenesis. It seems interesting that a high cholesterol diet improved spatial memory in experimental animals. However, the mechanisms by which cholesterol affects the memory still remains uncovered (Sparks and Schreurs 2003; Nelson and Alkon 2005). Other possible scenario by which cholesterol deficiency could impair the function of the brain cells should also be taken into account. The alterations of cholesterol content may change the fusibility of membranes, normal functioning of glutamate receptors as well as other membrane proteins involved in synaptic transmission, such as ion channels, pumps, and SNAREs (Subtil et al. 1999; Launikonis and Stephenson 2001; Mauch et al. 2001; Hill et al. 2002; Hering et al. 2003; Pfrieger 2003; Salaun et al. 2004, 2005; Lange et al. 2005; Rohrbough and Broadie 2005; Churchward et al. 2005; Wasser et al. 2007; Allen et al. 2007; Chattopadhyay and Paila 2007; Cho et al. 2007).

References

Allen JA, Halverson-Tamboli RA, Rasenick MM (2007) Lipid rafts microdomains and neurotransmitter signaling. Nat Rev Neurosci 8:128–140

Baker DA, Xi ZX, Shen H et al (2002) The origin and neuronal function of in vivo nonsynaptic glutamate. J Neurosci 22:9134–9141

Borisova T, Krisanova N, Sivko R, Borysov A (2010a) Cholesterol depletion attenuates tonic release but increases the ambient level of glutamate in rat brain synaptosomes. Neurochem Int 56:466–478

Borisova T, Sivko R, Borysov A, Krisanova N (2010b) Diverse presynaptic mechanisms underlying methyl-beta-cyclodextrin—mediated changes in glutamate transport. Cell Mol Neurobiol 30:1013–1023

Butchbach M, Tian G, Guo H et al (2004) Association of excitatory amino acid transporters, especially EAAT2, with cholesterol-rich lipid raft microdomains. J Biol Chem 279:34388–34396

Cavelier P, Attwell D (2005) Tonic release of glutamate by a DIDS-sensitive mechanism in rat hippocampal slices. J Physiol 564:397–410

Cavelier P, Hamann M, Rossi D, Mobbs P, Attwell D (2005) Tonic excitation and inhibition of neurons: ambient transmitter sources and computational consequences. Prog Biophys Mol Biol 87:3–16

Cebers G, Cebere A, Kovács AD, Högberg H, Moreira T, Liljequist S (2001) Increased ambient glutamate concentration alters the expression of NMDA receptor subunits in cerebellar granule neurons. Neurochem Int 39:151–160

Chattopadhyay A, Paila YD (2007) Lipid-protein interactions, regulation and dysfunction of brain cholesterol. Biochem Biophys Res Commun 354:627–633

Cho WJ, Jeremic A, Jin H et al (2007) Neuronal fusion pore assembly requires membrane cholesterol. Cell Biol Int 31:1301–1308

Churchward MA, Rogasevskaia T, Hofgen J et al (2005) Cholesterol facilitates the native mechanism of Ca²⁺-triggerted membrane fusion. J Cell Sci 118:4833–4848

Dalby NO, Mody I (2003) Activation of NMDA receptors in rat dentate gyrus granule cells by spontaneous and evoked transmitter release. J Neurophysiol 90:786–797

Forsythe ID, Clements JD (1990) Presynaptic glutamate receptors depress excitatory monosynaptic transmission between mouse hippocampal neurones. J Physiol 429:1–16

Frank C, Rufini S, Tancredi V, Forcina R et al (2008) Cholesterol depletion inhibits synaptic transmission and synaptic plasticity in rat hippocampus. Exp Neurol 212:407–414

Hering H, Lin CC, Sheng M (2003) Lipid rafts in the maintenance of synapses, dendritic spines, and surface AMPA receptor stability. J Neurosci 23:3262–3271

Hill W, An B, Johnson J (2002) Endogenously expressed epithelial sodium channel is present in lipid rafts in A6 cells. J Biol Chem 277:33541–33544

Jabaudon D, Shimamoto K, Yasuda-Kamatani Y (1999) Inhibition of uptake unmasks rapid extracellular turnover of glutamate of nonvesicular origin. Proc Natl Acad Sci USA 96:8733–8738

Karten B, Vance DE, Campenot RB et al (2002) Cholesterol accumulates in cell bodies, but is decreased in distal axons, of Niemann-Pick C1-deficient neurons. J Neurochem 83:1154–1163

Karten B, Vance DE, Campenot RB et al (2003) Trafficking of cholesterol from cell bodies to distal axons in Niemann Pick C1-deficient neurons. J Biol Chem 278:4168–4175

Kenet T, Bibitchkov D, Tsodyks M (2003) Spontaneously emerging cortical representations of visual attributes. Nature 425:954–956

Krisanova N, Sivko R, Kasatkina L, Borisova T (2012) Neuroprotection by lowering cholesterol: a decrease in membrane cholesterol content reduces transporter-mediated glutamate release from brain nerve terminals. Biochim Biophys Acta 1822:1013–1023

Lange Y, Ye J, Steck TL (2005) Activation of membrane cholesterol by displacement from phospholipids. J Biol Chem 280:36126–36131

Launikonis BS, Stephenson DG (2001) Effects of membrane cholesterol manipulation on excitation-contraction coupling in skeletal muscle of the toad. J Physiol 534:71–85

Mason RP (1994) Probing membrane bilayer interactions of 1,4-dihydropyridine calcium channel blockers. Implications for aging and Alzheimer's disease. Ann N Y Acad Sci 747:125–139

Mauch DH, Nägler K, Schumacher S et al (2001) CNS synaptogenesis promoted by glia-derived cholesterol. Science 294:1354–1357

McBain CJ, DiChiara TJ, Kauer JA (1994) Activation of metabotropic glutamate receptors differentially affects two classes of hippocampal interneurons and potentiates excitatory synaptic transmission. J Neurosci 14:4433–4445

Meur KL, Galante M, Angulo MC, Audinat E (2007) Tonic activation of NMDA receptors by ambient glutamate of non-synaptic origin in the rat hippocampus. J Physiol 580:373–383

Michikawa M (2003) The role of cholesterol in pathogenesis of Alzheimer's disease: dual metabolic interaction between amyloid beta-protein and cholesterol. Mol Neurobiol 27:1–12

Molander-Melin M, Blennow K, Bogdanovic N et al (2005) Structural membrane alterations in Alzheimer brains found to be associated with regional disease development; increased density of gangliosides GM1 and GM2 and loss of cholesterol in detergent-resistant membrane domains. J Neurochem 92:171–182

Nelson TJ, Alkon DL (2005) Insulin and cholesterol pathways in neuronal function, memory and neurodegeneration. Biochem Soc Trans 33:1033–1036

Nicoll JA, Roberts GW, Graham DI (1995) Apolipoprotein E epsilon 4 allele is associated with deposition of amyloid beta-protein following head injury. Nat Med 1:135–137

Nyitrai G, Kékesi KA, Juhász G (2006) Extracellular level of GABA and Glu: in vivo microdialysis-HPLC measurements. Curr Top Med Chem 6:935–940

Ohnishi M, Watanabe Y, Shibuya T (1995) Potentiation of excitotoxicity by glutamate uptake inhibitor rather than glutamine synthetase inhibitor. Jpn J Pharmacol 68:315–321

Patel AJ, Hunt A, Tahourdin CSM (1983) Regulation of in vivo glutamine synthetase activity by glucocorticoids in the developing rat brain. Dev Brain Res 10:83–91

Pfrieger FW (2003) Cholesterol homeostasis and function in neurons of the central nervous system. Cell Mol Life Sci 60:1158–1171

Ponce J, Pérez de la Ossa N, Hurtado O et al (2008) Simvastatin reduces the association of NMDA receptors to lipid rafts. A cholesterol-mediated effect in neuroprotection. Stroke 39:1269–1275

Roher AE, Weiss N, Kokjohn TA et al (2002) Increased A beta peptides and reduced cholesterol and myelin proteins characterize white matter degeneration in Alzheimer's disease. Biochemistry 41:11080–11090

Rohrbough J, Broadie K (2005) Lipid regulation of the synaptic vesicle cycle. Nat Rev Neurosci 6:139–150

Sah P, Hestrin S, Nicoll RA (1989) Tonic activation of NMDA receptors by ambient glutamate enhances excitability of neurons. Science 246:815–818

Salaun C, James DJ, Chamberlain LH (2004) Lipid rafts and the regulation of exocytosis. Traffic 5:1–10

Salaun C, Gould GW, Chamberlain LH (2005) Lipid raft association of SNARE proteins regulates exocytosis in PC12 cells. J Biol Chem 280:19449–19453

Sparks DL, Schreurs BG (2003) Insulin and cholesterol pathways in neuronal function, memory and neurodegeneration. Proc Natl Acad Sci USA 100:11065–11069

Subtil A, Gaidarov I, Kobylarz K et al (1999) Acute cholesterol depletion inhibits clathrin-coated pit budding. Proc Natl Acad Sci USA 96:6775–6780

Svennerholm L, Gottfries CG (1994) Membrane lipids, selectively diminished in Alzheimer brains, suggest synapse loss as a primary event in early-onset form (type I) and demyelination in late-onset form (type II). J Neurochem 62:1039–1047

Tansey FA, Farooq M, Cammer W (1991) Glutamine synthetase in oligodendrocytes and astrocytes: new biochemical and immunocytochemical evidence. J Neurochem 56:266–272

Teasdale GM, Nicoll JA, Murray G et al (1997) Association of apolipoprotein E polymorphism with outcome after head injury. Lancet 350:1069–1071

Tsai HI, Tsai LH, Chen MY et al (2006) Cholesterol deficiency perturbs actin signaling and glutamate homeostasis in hippocampal astrocytes. Brain Res 1104:27–38

Wasser CR, Ertunc M, Liu X et al (2007) Cholesterol-dependent balance between evoked and spontaneous vesicle recycling. J Physiol 579(2):413–429

Westphalen RI, Hemmings HC Jr (2006) Volatile anesthetic effects on glutamate versus GABA release from isolated rat cortical nerve terminals: basal release. J Pharmacol Exp Ther 316:208–215

Yao ZX, Papadopoulos V (2002) Function of beta-amyloid in cholesterol transport: a lead to neurotoxicity. FASEB J 16:1677–1679

Chapter 5
Unstimulated and Exocytotic Glutamate Release from Cholesterol-Deficient Nerve Terminals

Abstract Net tonic release of L-[^{14}C]glutamate from cholesterol-depleted synaptosomes was decreased by 38 % and release in low-Na$^+$ medium was attenuated by 41 % in the presence of DL-threo-β-benzyloxyaspartate, which significantly reduced glutamate uptake. It was suggested that cholesterol deficiency altered the intra-to-extracellular glutamate ratio by the reduction of the cytosolic level of the neurotransmitter and the augmentation of the ambient glutamate level, thereby favoring a decrease in tonic glutamate release. Increased extracellular glutamate in cholesterol-deficient nerve terminals was not a result of the changes in tonic release (and/or glutamine synthetase activity), but was set by lack function of glutamate transporters.

Cholesterol depletion with MβCD acutely applied to rat brain synaptosomes is accompanied by an immediate increase in transporter-mediated glutamate release and a decrease in exocytotic release.

5.1 Tonic (Unstimulated) Release of Glutamate from Cholesterol-Depleted Synaptosomes

As it was mentioned above, glutamate release via spontaneous exocytosis, swelling-activated anion channels, cystine-glutamate exchange and transmembrane diffusion are constituents of tonic, i.e., unstimulated, release of the neurotransmitter from the nerve terminals; however, the exact origin of this release has not been identified yet (Rutledge et al. 1998; Jabaudon et al. 1999; Cavelier and Attwell 2005). Baker et al. (2002) has suggested from microdialysis experiments that 60 % of tonic glutamate release in the striatum is generated by cystine–glutamate exchange (in which glutamate is released in exchange for cystine taken up to make glutathione). Blocking voltage-gated Ca^{2+} channels with 200 μm Cd^{2+}, and using bafilomycin pretreatment to block the vesicular H$^+$-ATPase (Rossi et al. 2003; Allen et al. 2004; Allen and Attwell 2004) in hippocampal slices, it has been shown that tonic release is not occurred via Ca^{2+}-dependent exocytosis (Jabaudon et al. 1999; Cavelier and Attwell 2005).

In AT, tonic release of L-[^{14}C]glutamate was significantly increased after the addition of MβCD to the nerve terminals that consisted of 0.001 ± 0.0001 nmol/min/ mg of protein in the control and 0.018 ± 0.001 nmol/min/mg of protein in the presence of 15 mM MβCD ($P \leq 0.001$, Student's t-test, $n=4$). It is clear that this increase in tonic release of L-[^{14}C]glutamate resulted from MβCD-evoked dissipation of the proton gradient of synaptic vesicles.

Basing on the equilibrium of the above chapter [extracellular glutamate] = [tonically released glutamate] – [glutamate uptaken by transporters], it was considered that net tonic release, the first term on the right side of this equation, meant unstimulated release, which was measured without the input of transporter-mediated uptake or at least with significantly reduced contribution of the latest one. Otherwise, if uptake is not attenuated, the released neurotransmitter is immediately deleted from the exrtracellular space by glutamate transporters, thereby apparently lowering the value of tonic release. The competitive non-transportable inhibitor of glutamate transporters DL-threo-β-benzyloxyaspartate (DL-TBOA) essentially inhibits the initial velocity of uptake of L-[^{14}C]glutamate from 3.0 ± 0.3 nmol $\times$ min^{-1} $\times$ mg^{-1} protein in the control up to 1.5 ± 0.2 nmol $\times$ min^{-1} $\times$ mg^{-1} protein during application of 10 μM DL-TBOA, 0.9 ± 0.1 nmol $\times$ min^{-1} $\times$ mg^{-1} protein in 100 μM DL-TBOA, and 0.63 ± 0.1 nmol $\times$ min^{-1} $\times$ mg^{-1} protein in 200 μM DL-TBOA ($P<0.01$, ANOVA, $n=6$). Thus, it seems accurate to apply 200 μM DL-TBOA for substantial inhibition of the activity of glutamate transporters in cholesterol deficiency (Borisova et al. 2010a).

The question asked was whether net tonic release of glutamate contributed to the augmentation of the extracellular glutamate level provoked by depletion of membrane cholesterol after LP. The application of DL-TBOA caused a different increase in the ambient level of preloaded L-[^{14}C]glutamate in control and cholesterol-depleted synaptosomes. The value of tonically released L-[^{14}C]glutamate during the application of the inhibitor (the resulting rise in ambient L-[^{14}C]glutamate calculated as a difference between 0 and 6 min/12 min time points) was decreased for 6 min from 7.8 ± 0.6 % of total label in the control to 4.9 ± 0.6 % of total label in MβCD-treated synaptosomes (LP) ($P \leq 0.05$, Student's t-test, $n=4$) and for 12 min from 10.8 ± 0.6 % to 8.87 ± 0.6 %, respectively ($P \leq 0.05$, Student's t-test, $n=4$) (Fig. 5.1). It should be noted that continuous 30 min monitoring also revealed a reduction in the value of tonic glutamate release in cholesterol-depleted synaptosomes as compared to the control. The next experiments were focused on the comparison of untreated control synaptosomes and the synaptosomes treated with 15 mM MβCD complexed with cholesterol (2.3 mM). No significant changes in the value of net tonic release of L-[^{14}C]glutamate in the presence of DL-TBOA were observed. Thus, net tonic glutamate release evaluated under conditions of the reduced input of glutamate transporters became lesser after cholesterol depletion. The attenuation of tonic glutamate release in cholesterol deficiency could be at least partially a result of an increase in the ambient glutamate level favoring a reduction of the glutamate gradient across the plasma membrane (Borisova et al. 2010a).

Glutamate dehydrogenase assay was also used for the assessment of tonic release of endogenous glutamate in cholesterol-deficient nerve terminals. Tonic release of

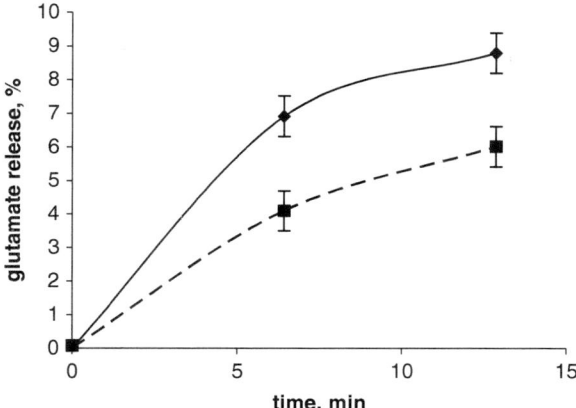

Fig. 5.1 Tonic release of preloaded L-[^{14}C]glutamate from control (*solid line*) and cholesterol-depleted synaptosomes (LP) (*dotted line*) during application of 200 μM DL-TBOA. The synaptosomes were incubated without MβCD (the control experiments), with 15 mM MβCD (LP, extraction of cholesterol) or 15 mM MβCD complexed with 2.3 mM cholesterol (analysis of the effects of MβCD irrespective to cholesterol accepting capacity) at 37 °C during 30 min followed by washing. Control, MβCD- and MβCD/cholesterol-treated synaptosomes were loaded with L-[^{14}C]glutamic acid (500 nM, 238 mCi/mmol) in Ca^{2+}-supplemented oxygenated standard salt solution. After loading, tonic release of L-[^{14}C]glutamate was measured in Ca^{2+}-free incubation media according to following method: samples (125 μL of the suspension, 0.5 mg of protein/mL) were preincubated for 8 min at 37 °C, then incubated further for 0–12 min for recording of tonic L-[^{14}C]glutamate release and rapidly sedimented in a microcentrifuge. L-[^{14}C]glutamate radioactivity in the supernatants was determined. Total synaptosomal L-[^{14}C]glutamate content was equal to 200,000±15,000 cpm/mg protein. Data are means±SEM of four independent experiments, each performed in triplicate. Data are compared by Student's *t*-test. Figure as in Borisova et al. (2010a)

endogenous glutamate from the synaptosomes was measured starting from 3 min time point when the most of ambient glutamate was converted to α-ketoglutarate by glutamate dehydrogenase. In cholesterol-deficient nerve terminals, tonic release of endogenous glutamate was decreased by ~25 % at 1 min time point in comparison with the control (Fig. 5.2). The application of 15 mM MβCD complexed with cholesterol (2.3 mM) caused insignificant changes in tonic release of glutamate from the synaptosomes as compared to the control. Thus, the results of the assessment of tonic release of glutamate from cholesterol-deficient synaptosomes obtained with glutamate dehydrogenase assay were in accordance with data on radiolabeled glutamate (Krisanova et al. 2012).

An elevation in the extracellular glutamate level was accompanied with an attenuation of tonic glutamate release because of the reduced glutamate gradient across the plasma membrane, where the intracellular glutamate level (glutamate in) should also merit consideration. This suggestion was true if the cytosolic glutamate level was not augmented or at least remained unchanged after depletion of membrane cholesterol. The application of digitonin, a specific detergent, which at low concentrations has been used for the plasma membrane permeabilization (Holz and Senter 1985; Katz and Wals 1985), allowed examining more precisely the cytosolic glutamate content (Krisanova et al. 2009). Digitonin evoked an immediate leakage of

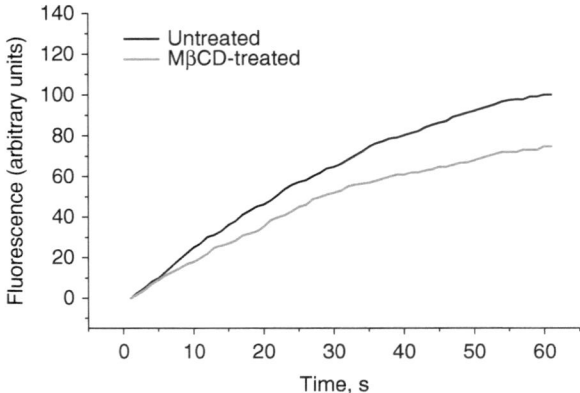

Fig. 5.2 Tonic release of endogenous glutamate from control (*black line*) and cholesterol-deficient (*grey line*) rat brain synaptosomes assessed with glutamate dehydrogenase assay. Control and 15 mM MβCD-treated synaptosomal suspension (0.5 mg/mL of final protein concentration) was added to an enzymatic assay solution containing glutamate dehydrogenase. Tonic release (starting from 3 min time point) of endogenous glutamate in the synaptosomes was measured by the changes in NADH fluorescence (excitation and emission wavelengths of 340 and 460 nm, respectively). Trace is representative of three independent experiments. Figure as in Krisanova et al. (2012)

cytosolic glutamate from the synaptosomes just after its application. This efflux occurred through the pores formed by digitonin, thereby eliminating the contribution of glutamate transporters to the estimated parameter. 15 μM digitonin caused a quick increase in the extracellular L-[^{14}C]glutamate level, which was registered at 1 min time point of digitonin application. Leakage of preloaded L-[^{14}C]glutamate was equal to 9.4 ± 0.4 % of total label in control and 7.8 ± 0.4 % of total label in cholesterol-depleted synaptosomes ($P \leq 0.05$, Student's t-test, $n = 4$), whereas in the synaptosomes treated with MβCD complexed with cholesterol it did not differ significantly from the control (Fig. 5.3). Thus, the cytosolic glutamate level became lower after depletion of membrane cholesterol as compared to the control. These data allowed to consider that weak uptake by cholesterol-deficient synaptosomes caused a decrease in the cytosolic level of glutamate. The lowered level of cytosolic glutamate reduced further the glutamate gradient across the plasma membrane in cholesterol deficiency and together with the increased extracellular content of the neurotransmitter caused a decrease in tonic release (Borisova et al. 2010a).

5.2 Influence of the Continuous Enrichment of the Cytoplasm with Vesicular Glutamate on Tonic Release of the Neuromediator

It is stoichiometrically predictable that the greater intracellular glutamate level, the greater net tonic release. The experimental conditions, which provided the continuous replenishment of the cytoplasm with the neurotransmitter originating from synaptic

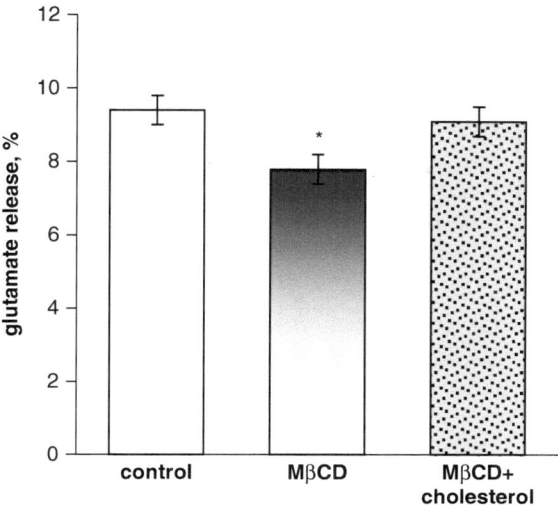

Fig. 5.3 Leakage of cytosolic L-[^{14}C]glutamate from digitonin-permeabilized synaptosomes in the control (*empty bar*), after cholesterol depletion (*shaded bar*) and the synaptosomes treated with MβCD complexed with cholesterol (*grey dotted bar*). The synaptosomes were incubated without MβCD (the control experiments), with 15 mM MβCD (LP, extraction of cholesterol) or 15 mM MβCD complexed with 2.3 mM cholesterol (analysis of the effects of MβCD irrespective to cholesterol accepting capacity) at 37 °C during 30 min followed by washing. Control, MβCD- and MβCD/cholesterol-treated synaptosomes were loaded with L-[^{14}C]glutamic acid (500 nM, 238 mCi/mmol) in Ca^{2+}-supplemented oxygenated standard salt solution. After loading, samples (125 µL of the suspension, 0.5 mg of protein/mL) were preincubated for 8 min at 37 °C, and then the synaptosomes were incubated with 15 µM digitonin for 1 min at 37 °C and rapidly sedimented in a microcentrifuge. The results were presented with basal glutamate release subtracted. L-[^{14}C] glutamate radioactivity in the supernatants was determined. Total synaptosomal L-[^{14}C]glutamate content was equal to 200,000 ± 15,000 cpm/mg protein. Data are means ± SEM of four independent experiments, each performed in triplicate. Data are compared by Student's *t*-test. *$P \leq 0.05$ as compared to the control or the synaptosomes treated with MβCD complexed with cholesterol. Figure as in Borisova et al. (2010a)

vesicles, should favor an augmentation of tonic glutamate release despite the enhancement of its extracellular level. A principal assumption in the use of the protonophore carbonyl cyanide-*p*-trifluoromethoxyphenyl-hydrazon (FCCP) is the ability to dissipate the proton gradient and inhibit uptake of glutamate in synaptic vesicles (Cidon and Sihra 1989; Tretter et al. 1998; Zoccarato et al. 1999). After application of FCCP, synaptic vesicles were not able to keep glutamate inside and the neurotransmitter was released into the cytosol. The evaluation of the above hypothesis was carried out on control synaptosomes without the treatment with MβCD. The cytosolic glutamate level was considerably enhanced during application of 1 µM FCCP that was registered by an enlargement of depolarization-stimulated transporter-mediated glutamate release for 6 min from 12.2 ± 0.8 % of total label to 23.2 ± 1.0 % of total label ($P \leq 0.05$, Student's *t*-test, $n = 8$). The extracellular glutamate level was increased twice from 0.193 ± 0.013 nmol/mg protein in the control to 0.400 ± 0.020 nmol/mg protein in the synaptosomes treated with

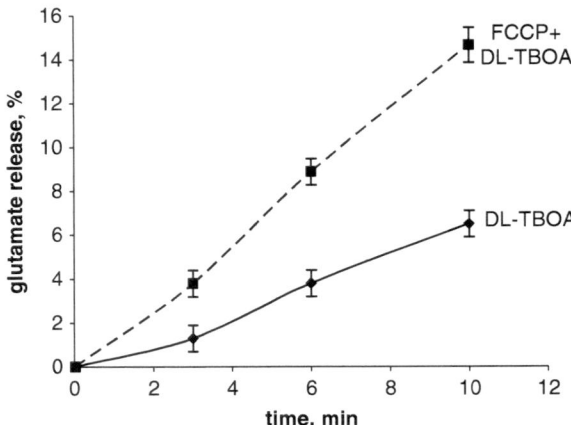

Fig. 5.4 Tonic release of preloaded L-[14C]glutamate in 200 μM DL-TBOA-containing, Ca^{2+}-free incubation media in the presence of the protonophore FCCP (*dotted line*) or without FCCP (*solid line*). The synaptosomes were loaded with L-[14C]glutamate (500 nM, 238 mCi/mmol). After loading, tonic release of L-[14C]glutamate was measured according to following method: samples (125 μL of the suspension, 0.5 mg of protein/mL) were preincubated for 8 min at 37 °C, then FCCP was added to the incubation medium and the synaptosomes were incubated for 0–10 min and rapidly sedimented in a microcentrifuge. L-[14C]glutamate radioactivity in the supernatants was determined. Total synaptosomal L-[14C]glutamate content was equal to 200,000±15,000 cpm/mg protein. (Tonic release in the presence of 200 μM DL-TBOA without the protonophore was lesser than it was shown in Fig. 5.1, because in these series of the experiments, the synaptosomes were not undergone 30 min preliminary incubation for considering them as a control for MβCD-treated synaptosomes). Data are means±SEM of four independent experiments, each performed in triplicate. Data are compared by Student's *t*-test. Figure as in Borisova et al. (2010a)

FCCP ($P \leq 0.05$, Student's *t*-test, $n = 8$). This rise in the ambient glutamate level resulted from an attenuation of glutamate uptake (by ~60 %) and an augmentation of transporter-mediated release of glutamate (by ~90 %) in the presence of the protonophore. To reduce the contribution of glutamate transporters working in both directions, i.e., uptake and release, 200 μM DL-TBOA, which is able to inhibit their direct and reversed work, was applied. This experimental protocol allowed to register net tonic release, when the cytosolic glutamate level has been continuously enlarging with vesicular glutamate during the application of FCCP. As it was clearly shown in Fig. 5.4, net tonic release of glutamate was increased twice for 6 min and consisted of 3.8±0.6 % of total label in the synaptosomes with DL-TBOA only (without FCCP) and 8.9±0.6 % of total label in the presence of DL-TBOA and FCCP ($P \leq 0.05$, Student's *t*-test, $n = 4$); for 10 min 6.5±0.6 % and 14.6±0.8 %, respectively, ($P \leq 0.05$, Student's *t*-test, $n = 4$). Thus, the permanently elevated level of cytosolic glutamate caused an increase in net tonic release because of the continuous augmentation of the intra-to-extracellular glutamate ratio. Moreover, in this case, net tonic release was increased despite the significantly enhanced level of ambient glutamate (Borisova et al. 2010a).

Synaptosomes

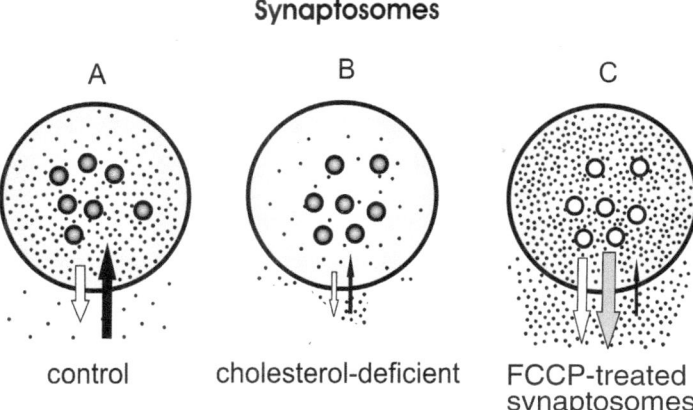

Fig. 5.5 The possible interrelation between uptake (*black arrows*), tonic release (*open arrows*), transporter-mediated release (*grey arrow*), and the intra/extra-cellular concentration of glutamate (indicated by *black dots*). Circles inside of the synaptosomes showed filled with the neurotransmitter (*black circles*) or empty (*open circles*) synaptic vesicles. Figure as in Borisova et al. (2010a)

The important observation was a decrease in tonic release and the cytosolic content of glutamate in cholesterol-deficient synaptosomes, in contrast, the extra-cellular level of the neurotransmitter was increased under these conditions. As it has been already mentioned, the origin of ambient glutamate is not completely identi-fied. It was suggested that glutamate enriched the extracellular space not only by spontaneous exocytosis, but also through swelling-activated anion channels, cystine-glutamate exchange, transmembrane diffusion, volume-sensitive Cl⁻ chan-nels, and possibly glutamate transporter reversal (Jabaudon et al. 1999; Rutledge et al. 1998; Cavelier and Attwell 2005). At least four of these components appear to be dependent from the intra-to-extra-cellular glutamate gradient. Wasser et al. (2007) demonstrated that the rate of spontaneous vesicle fusion was significantly increased in MβCD-treated hippocampal neurons; however, the contribution of spontaneous exocytosis as one of the components determining total tonic release should be assessed. It was supposed that the changes in ambient glutamate resulted from an inhibition of glutamate transporter activity in cholesterol-deficient synapto-somes, whereas the attenuation of tonic release was due to a decrease in the ratio (glutamate in)/(glutamate out) (Fig. 5.5a, b). It was confirmed experimentally that a continuous increase in tonic release occurred, when vesicular glutamate was con-stantly replenished the cytosol (Fig. 5.4). In this case, the intracellular glutamate concentration was augmented because of the inability of synaptic vesicles to keep the neuromediator. It is so because glutamate was not kept inside of synaptic vesi-cles, thereby enriching the cytosol. Whereas, under conditions of the absence of vesicular glutamate leakage, weak transporter-mediated uptake and the increased extracellular level of glutamate attenuated tonic release of the neurotransmitter (Figs. 5.1 and 5.2). In this context, the second interesting finding was the observation that under conditions of energy deprivation, at least partial dissipation of the proton gradient of synaptic vesicles could favor an augmentation of tonic glutamate release due to the continuously increasing ratio (glutamate in)/(glutamate out) (Fig. 5.5c). It should be noted that not only tonic release but also transporter-mediated release of glutamate was elevated during dissipation of the synaptic vesicle proton gradient (Borisova et al. 2010a).

The fact that cholesterol deficiency attenuated tonic release showed that the energetic status of the synaptosomes did not change after cholesterol depletion. This may be considered in support of the data of Barnes et al. (2004) concerning the unchanged level of ATP in the rat epithelial Clone 9 cell line after cholesterol depletion.

The cellular mechanisms, by which cholesterol could be involved in the development of the pathological processes in the nervous system, were analyzed. The results indicated that the enlargement of the extracellular level of glutamate in cholesterol-deficient nerve terminals resulted from the inhibition of glutamate transporter functioning, whereas the changes in tonic glutamate release and/or glutamine synthetase activity did not lead to an elevation in the ambient glutamate concentration. Thus, modulation of glutamate transporter activity by manipulation of membrane cholesterol is a possible therapeutic strategy in neurodegenerative diseases. The experimental data support the suggestion that modification of the neuronal plasma membrane, i.e., its cholesterol content, should be considered as a possible practical approach for the treatment of the neurological symptoms, compensation and prevention of the development of pathogenic mechanisms. The current observations suggested that cholesterol as an endogenous modulator of neurotransmission in the CNS could play a significant role in the neuroprotection and also might be extremely important for the synaptic plasticity (Borisova et al. 2010a).

5.3   Leakage of Glutamate from Cholesterol-Deficient Nerve Terminals at Low Temperature

The extracellular level and leakage of preloaded L-[^{14}C]glutamate from control and cholesterol-deficient synaptosomes after LP were analyzed at low temperature (+4 °C). Control and cholesterol-deficient synaptosomes were loaded with L-[^{14}C] glutamate at +37 °C, and then cooled to +4 °C. In the nerve terminals at 37 °C, the extracellular level of glutamate is determined by a balance between tonic release and uptake of the neurotransmitter. When the synaptosomes were cooled to +4 °C, the efficiency of all processes that usually established the extracellular glutamate concentration was drastically decreased.

The extracellular level of L-[^{14}C]glutamate in the synaptosomes at +4 °C consisted of 0.193 ± 0.013 nmol/mg of protein in control experiments and 0.410 ± 0.015 nmol/mg of protein after cholesterol depletion with 15 mM MβCD ($P \leq 0.05$, Student's t-test, $n = 5$) (Fig. 5.6). For the first 15 min of incubation of the synaptosomes at +4 °C, leakage of preloaded L-[^{14}C]glutamate from cholesterol-depleted synaptosomes was approximately 3 times higher in comparison with control ones and consisted of

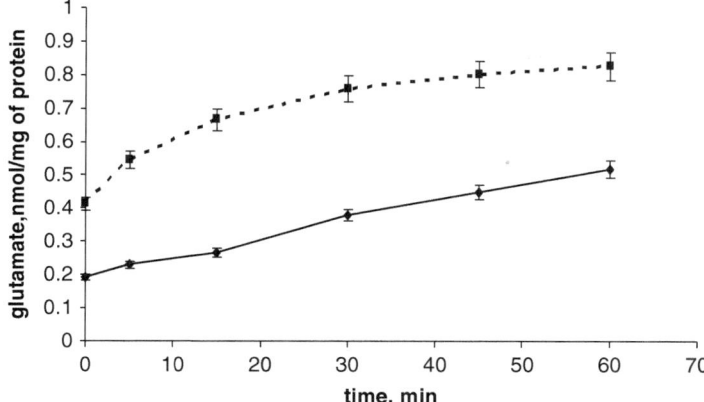

Fig. 5.6 The extracellular level of L-[^{14}C]glutamate in rat brain nerve terminals of normal and decreased level of cholesterol at +4 °C. Cholesterol extraction was performed using MβCD in the concentration of 15 mM. Figure as in Sivko et al. (2012)

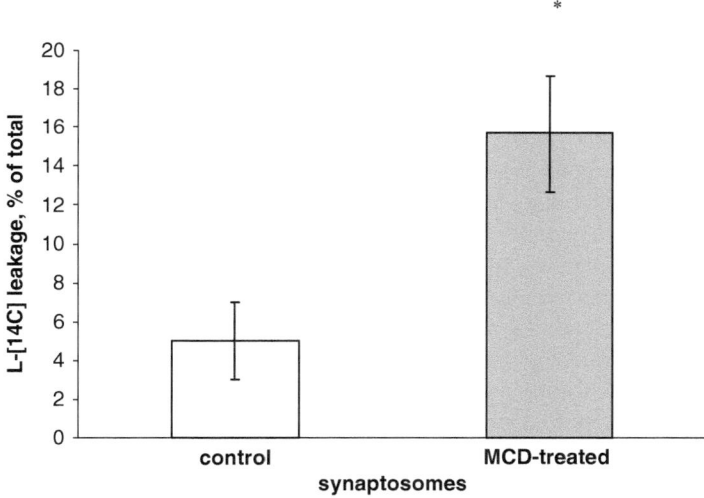

Fig. 5.7 Leakage of preloaded L-[^{14}C]glutamate from rat brain nerve terminals of normal and decreased level of cholesterol at +4 °C. Cholesterol extraction was performed using MβCD in the concentration of 15 mM. *$P \le 0.05$ as compared to the control. Figure as in Sivko et al. (2012)

5.0 ± 2 % of total label accumulated by the synaptosomes in control experiments and 15.7 ± 3 % of total label accumulated by synaptosomes after depletion of cholesterol ($P \le 0.05$, Student's t-test, $n = 5$) (Fig. 5.7). During further 60 min of incubation of synaptosomes at +4 °C, the difference in leakage of L-[^{14}C]glutamate between control and cholesterol-deficient synaptosomes became lesser, but the extracellular level of

L-[^{14}C]glutamate in cholesterol-deficient synaptosomes continued to be higher than that in control ones (Sivko et al. 2012).

At +4 °C, the synaptosomes preliminary treated with this complex MβCD/cholesterol, i.e., "neutral complex," 15 mM MβCD and 2.3 mM cholesterol, exhibited only insignificant changes in the extracellular L-[^{14}C]glutamate concentration as compared to untreated synaptosomes.

One of the reasons (except different membrane permeability of the synaptosomes for L-[^{14}C]glutamate) for an increase in leakage of L-[^{14}C]glutamate from cholesterol-deficient synaptosomes cooled to +4 °C may be the enlargement of the glutamate concentration in the cytosol. However, detergent digitonin, which is used for permeabilization of the plasma membrane, evoked (at +37 °C) lesser leakage of L-[^{14}C]glutamate (by ~17 %) from cholesterol-deficient nerve terminals in comparison with the control. The synaptosomes were warmed to +37 °C and monitored the ambient level of L-[^{14}C]glutamate. The difference between control and cholesterol-deficient synaptosomes became lesser and consisted of 0.193 ± 0.013 nmol/mg of protein in control experiments and 0.282 ± 0.013 nmol/mg of protein after cholesterol depletion with 15 mM MβCD ($P \leq 0.05$, Student's t-test, $n = 5$). Thus, the effect of cholesterol depletion on the extracellular level of L-[^{14}C]glutamate at +4 and +37 °C was similar.

At +37 °C tonic release of L-[^{14}C]glutamate from cholesterol-depleted nerve terminals was decreased in comparison with the control. However, at +4 °C, leakage of L-[^{14}C]glutamate from cholesterol-depleted synaptosomes was significantly increased in comparison with the control (Fig. 5.7). Thus, the effect of cholesterol depletion on leakage of L-[^{14}C]glutamate from the synaptosomes at +4 and +37 °C was opposite.

In the synaptosomes at +4 and +37 °C, the mechanisms underlying leakage of L-[^{14}C]glutamate and its extracellular level are different. In cooled synaptosomes, these parameters depend from membrane permeability only. Whereas, after warming of the synaptosomes, all mechanisms involved into the maintenance of the extracellular glutamate concentration were renewed, e.g., spontaneous exocytosis, functioning of glutamate transporter, anion channels, and cystine-glutamate exchanger. It should be mentioned that after warming of the synaptosomes the difference in the extracellular level between normal and cholesterol-deficient nerve terminals was retained.

The microviscosity of the lipids of the plasma membranes isolated from brain cells and endoplasmic reticulum membranes was increased within the entire temperature range from +10 to +45 °C that was registered by the time of rotation correlation of 2,2,6,6-tetramethyl-4-capryloyloxypyperidine-1-oxyl (Maltseva et al. 1991, 1996). Modulation in lipid composition of membranes is considered as a contributing factor in the ability of membranes to undergo modification in morphology such as a formation of hexagonal phase. Cholesterol regulates lipid chain order underlying many of membrane properties including movement of permeants.

Thus, at low temperature the membrane permeability of the nerve terminals to glutamate was higher after cholesterol depletion. Modulation of cholesterol level could influence temperature-induced changes in the properties of membranes (Sivko et al. 2012).

5.4   Exocytotic Release of Glutamate During Acute and Long-Term Treatment of the Synaptosomes with MβCD

Ca^{2+}-dependent exocytotic release of L-[^{14}C]glutamate from the synaptosomes was investigated just after the application of MβCD. Release was stimulated with high KCl in the presence of Ca^{2+} in the medium. Depolarization of the plasma membrane in the presence of Ca^{2+} is known to result in release of the neurotransmitter from both pools, vesicular and cytoplasmic, whereas depolarization in Ca^{2+}-free medium causes the entrance of Na^+ through potential-dependent sodium channels and stimulates transporter-mediated release of glutamate from the cytoplasm. Figures 5.8 and 5.9

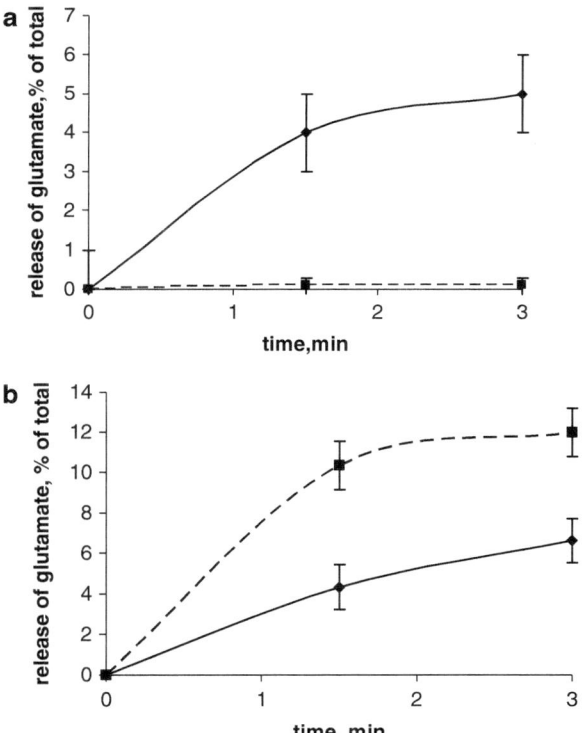

Fig. 5.8 Reduction of Ca^{2+}-dependent (exocytotic) release (**a**) and enhancement of Ca^{2+}-independent (transporter-mediated) release (**b**) of glutamate stimulated by high KCl during cholesterol depletion (AT): *solid line*—control synaptosomes; *dotted line*—synaptosomes in the presence of 5 mM MβCD, which was added at zero time point just before the application of 35 mM KCl. At 1.5- and 3-min time point aliquots of the samples were centrifuged and L-[^{14}C]glutamate radioactivity in the supernatants was determined. The amount of radiocarbon released into the buffer was expressed as percentage of total radioactivity in the synaptosomes. Total synaptosomal L-[^{14}C] glutamate content was equal to 200,000 ± 15,000 cpm/mg protein. Data are means ± SEM of five independent experiments, each performed in triplicate. Figure as in Tarasenko et al. (2010)

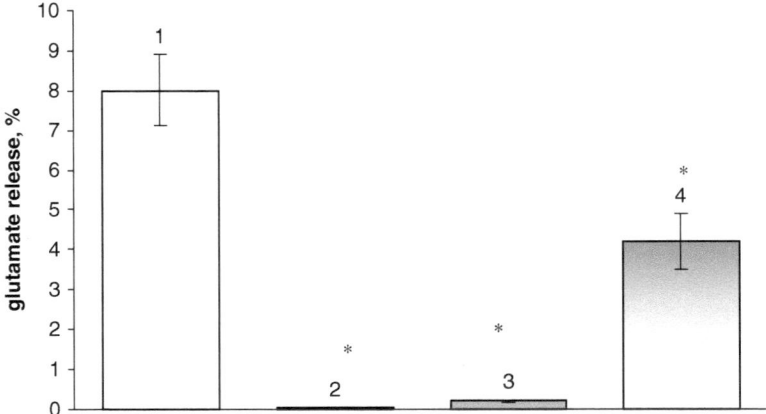

Fig. 5.9 Ca^{2+}-dependent (exocytotic) release of glutamate stimulated by high KCl during cholesterol depletion: 1—control synaptosomes; 2—the synaptosomes in the presence of 15 mM MβCD, which was added at zero time point just before application of 35 mM KCl; 3—the synaptosomes in the presence of 15 mM MβCD, which was added 35 min before the application of 35 mM KCl; 4—the synaptosomes were preliminary treated with 15 mM MβCD for 35 min, then were washed with 10 volumes of buffer, and then were loaded with L-[^{14}C]glutamate. At 6 min time point, the aliquots of the samples were centrifuged and L-[^{14}C]glutamate radioactivity in the supernatants was determined. The amount of radiocarbon released into the buffer was expressed as percentage of total radioactivity in the synaptosomes. Total synaptosomal L-[^{14}C]glutamate content was equal to $200,000 \pm 15,000$ cpm/mg protein. Data are means$\pm$SEM of five independent experiments, each performed in triplicate. *$P \leq 0.05$ as compared to the control. Figure as in Borisova et al. (2010b)

(the first and second columns) shows that cholesterol depletion of the synaptosomes with 5 mM MβCD is accompanied by a significant reduction (up to complete inhibition) of exocytotic L-[^{14}C]glutamate release. Such drastic effect of MβCD on exocytotic release of glutamate was not observed, when the acceptor was applied to the synaptosomes in complex with cholesterol (15 mM MβCD and 2.3 mM cholesterol) to prevent plasma membrane cholesterol depletion (Borisova et al. 2010b; Tarasenko et al. 2010).

The incubation of the synaptosomes at 37 °C for 35 min in the presence of the cholesterol acceptor only partially restored Ca^{2+}-dependent release of L-[^{14}C]glutamate (Fig. 5.9, the third column), whereas when this procedure followed by washing of MβCD from the incubation media the value of exocytotic L-[^{14}C]glutamate release decreased only by ~50 % and was equal to 8 ± 0.9 % of total in the control and 4.2 ± 0.7 % of total in cholesterol deficiency (Fig. 5.9, the first and fourth columns) ($P \leq 0.05$, Student's t-test, $n=5$). Preliminary data showed that the washing procedure per se did not change significantly the value of exocytotic L-[^{14}C]glutamate release from the synaptosomes. The synaptosomes were also treated with 15 mM MβCD complexed with cholesterol (2.3 mM). MβCD-cholesterol complex caused insignificant changes in exocytotic L-[^{14}C]glutamate release from the synaptosomes during acute and long-term treatment as compared to the control (Borisova et al. 2010b).

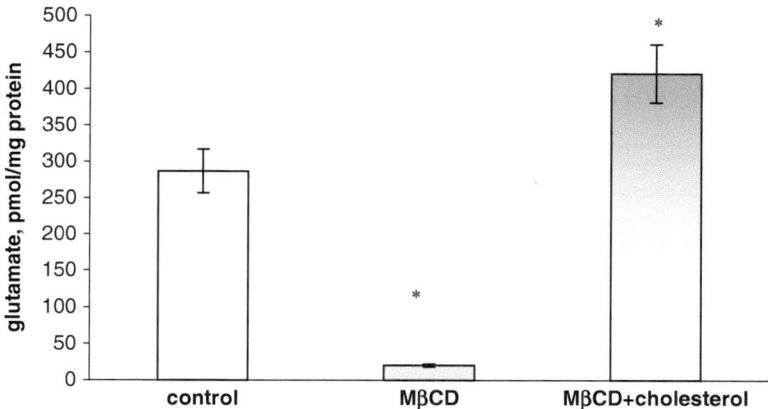

Fig. 5.10 The initial velocity of L-[^{14}C]glutamate (50 μM) uptake by isolated synaptic vesicles during application of 15 mM MβCD and 15 mM MβCD complexed with cholesterol (2.3 mM). Uptake of L-[^{14}C]glutamate by synaptic vesicles was measured as follows: samples (125 μL of the suspension, 1 mg of protein/mL) were pre-incubated in standard salt solution for 10 min at 30 °C. Uptake was initiated by the addition of 50 μM L-glutamate supplemented with 420 nM L-[^{14}C]gluta-mate (0.1 μCi/mL), incubated for 0–10 min at 30 °C. Uptake was measured in aliquots of patterns (100 μL) by filtration through the Millipore filters HAWP 025 00 washing with 4 mL warm buffer containing 5 mM Hepes-Tris, pH 7.4, 260 mM sucrose, and also in the aliquots of the supernatant (100 μL) and pellets rapidly sedimented in a microcentrifuge (20 s at 10,000 × g) by liquid scintilla-tion counting with scintillation cocktail OSC for the filters and ACS for the supernatants and pellets (1.5 mL). Nonspecific binding of the neurotransmitter was evaluated in the absence of ATP in the incubation media. Results were expressed as mean ± SEM values. Data are means ± SEM of three independent experiments, each performed in triplicate. *$P \leq 0.05$ as compared to the control. Figure as in Borisova et al. (2010b)

High sensitivity of L-[^{14}C]glutamate uptake to the level of cholesterol in vesicle membrane was demonstrated in the experiments with isolated synaptic vesicles. As seen in Fig. 5.10, MβCD (15 mM) itself caused a significant decrease in the initial velocity of L-[^{14}C]glutamate uptake in isolated synaptic vesicles as compared to the control, whereas MβCD complexed with cholesterol had the opposite effect (Borisova et al. 2010b).

Lang et al. (2001) in studies of exocytotic release of dopamine from PC12 cells and Gil et al. (2005), who investigated high K$^+$- or Ca^{2+} ionophore-stimulated release of glutamate from rat brain synaptosomes, showed a considerable decrease in exocytosis. However, in spite of numerous studies, there is no clear concept about the mechanisms underlying the modulation of exocytotic process under cholesterol depletion. Perhaps, the impairment of exocytosis is a consequence of partial inacti-vation of plasma membrane proteins involved in exocytosis, the most of them (voltage-gated Ca^{2+} and K$^+$ channels, syntaxin and SNAP-25) are embedded in cholesterol-rich microdomains (Lang et al. 2001; Salaun et al. 2004; Taverna et al. 2004; Xia et al. 2004, 2007). The changes in exocytotic release during acute and long-term treatment without washing of the acceptor were associated with the lower

Table 5.1 The table summarizing data on the changes in the key characteristics of synaptic neurotransmission during AT and LT with 15 mM MβCD

Synaptic neurotransmission characteristics	Control, %	AT with MβCD; the changes in comparison with the control, %	LT with MβCD; the changes in comparison with the control, %
The ambient glutamate level	100	↑ ~200	↑ ~130
Tonic glutamate release	100	↑ ~1,500	↓ ~70
Exocytotic release of glutamate	100	↓ 0	↓ ~50
Transporter-mediated release of glutamate	100	↑ ~200	↓ ~60
Glutamate uptake	100	↓ ~40	↓ ~60

Table as in Borisova et al. (2010b)

proton gradient of synaptic vesicles, which were not able to keep glutamate inside, thereby decreasing the value of exocytotic release. As Taverna et al. (2004) showed that potential-dependent calcium channels were located in lipid rafts, it seemed plausible to suggest that their activity was also changed after LT. The experiments with the calcium ionophores revealed that release stimulated by ionomycin (1 μM) or A 23187 (5 μM) was decreased in the synaptosomes after LT from 20.0 ± 1.9 % of total to 14.0 ± 1.9 % of total ($P \leq 0.05$, $n = 12$) and from 21.3 ± 1.9 % of total to 4.6 ± 1.9 % of total ($P \leq 0.05$, $n = 9$), respectively. Thus, the changes in exocytotic glutamate release after LT were not related to the alterations in calcium influx through potential-sensitive calcium channels. After LT, when the acceptor was deleted from the incubation media, the acidification of synaptic vesicles was restored and the changes in Ca^{2+}-dependent glutamate release seemed to be caused by alterations in the exocytotic machinery (Borisova et al. 2010b).

We reported that different alterations were found in exocytotic, transporter-mediated, tonic release and uptake of glutamate as well as its ambient level in cholesterol-deficient synaptosomes after AT and LT by MβCD (Table 5.1; Fig. 5.11). Moreover, we have shown that these changes were associated with disturbance of different presynaptic mechanisms in the nerve terminals (Borisova et al. 2010b) (Table 5.1).

Nevertheless, by assuming that MβCD action is realized only at the level of the plasma membrane, it is difficult to explain the significant enhancement of cytosolic glutamate concentration observed. The results suggested that MβCD-induced cholesterol extraction could have a strong impact on the ability of synaptic vesicles to keep glutamate inside, causing the redistribution of the neurotransmitter between the vesicles and cytoplasm. In this context, we hypothesized that the main cause of this phenomenon was a disturbance of the electrochemical H^+ gradient across the synaptic vesicle membrane, which is a driving force for neurotransmitter accumulation into synaptic vesicles. Furthermore, it has been demonstrated that vesicular proton pump (V type H^+-ATP-ase), which generates the electrochemical proton gradient across the synaptic vesicle membrane, is a cholesterol-binding protein, and interaction of this H^+-ATP-ase with cholesterol is important for its functional activity

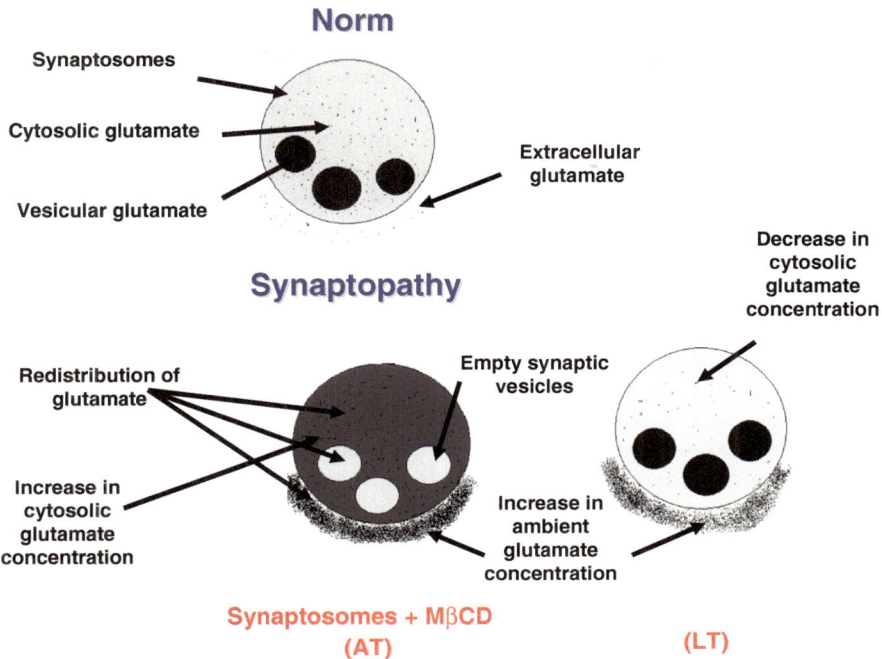

Fig. 5.11 Glutamate redistribution in cholesterol-deficient synaptosomes after AT and LT with MβCD

(Thiele et al. 2000). Omission of cholesterol in phospholipid vesicles was shown to inhibit the pumping activity of reconstituted vacuolar H$^+$-ATP-ase, thereby affecting the development of ΔpH (Perez-Castiñeira and Apps 1990). In this respect, the evidence that extracellularly applied MβCD can extensively deplete cholesterol from the intracellular membrane compartments (Jadot et al. 2001; Lange et al. 2004) allows to hypothesize that MβCD-evoked effects may be also due to cholesterol depletion from the synaptic vesicle membranes, since cholesterol is their prominent structural component (Deutsch and Kelly 1981).

Depletion of plasma membrane cholesterol with MβCD corresponds to dose-dependent leakage of the dye (and so protons) from synaptic vesicles and this process is not enhanced by extracellular Ca^{2+}. Confocal laser scanning microscopy and spectrofluorometric studies, which were performed using pH-sensitive fluorescent dye acridine orange, have shown that MβCD acutely applied to the synaptosomes immediately impacted the functional state of synaptic vesicles (Tarasenko et al. 2010). These observations are significant and together with an essential increase in transporter-mediated glutamate release suggest that massive release of the dye is attributable to dissipation of the proton gradient across the synaptic vesicle membrane rather than to exocytosis. The electron microscopy data of Wasser et al. (2007) is in favor of this assumption, demonstrating that the total number of synaptic vesicles per synapse in cultures treated with MβCD is reduced only by 8 %. Furthermore, it is known that the plecomacrolides, bafilomycin and concanamycin, the inhibitors of vesicular H$^+$-ATPase, cause leakage of both protons and glutamate from the vesicles (Zhou et al. 2000) as a result of the interaction with cholesterol-binding Vo subunit c of this enzyme (Huss and Wieczorek 2009; Zhang et al. 1994). Therefore, dissipation of synaptic vesicle proton gradient and a significant increase in glutamate release from the cytosol monitored just after the addition of MβCD to synaptosomal suspension might be a consequence of V-type H$^+$-ATPase dysfunction attributable to cholesterol depletion from the membrane of synaptic vesicles. From this point of view, a drastic decrease in exocytotic release of glutamate from the nerve terminals after the addition of MβCD could be explained by the involvement of incompletely filled vesicles in exocytosis (Tarasenko et al. 2010).

One of the main uncertainties of this study is that only indirect evidence of disturbed cholesterol content of synaptic vesicles during acute application of MβCD to the synaptosomes can be presented. From one side, the direct measurement of synaptic vesicle cholesterol content in lysed synaptosomes is incorrect because of possible influence of MβCD presenting in the incubation media on synaptic vesicles. From other side, washing of the synaptosomes from MβCD before lyses needs the time that appears to be enough for cholesterol redistribution and restoration (Steck et al. 2002; Yancey et al. 1996). This suggestion was supported by the data, which revealed the absence of dramatic changes in the proton gradient of synaptic vesicles in washed synaptosomes. It may also be speculated that drastic dissipation of synaptic vesicle proton gradient is associated with the falling electrochemical gradients across the plasma membrane during acute extraction of cholesterol. However, the experiments with the potentiometric optical dye rhodamine 6G showed that MβCD acutely applied to the nerve terminals caused only minor depolarization of the plasma membrane (Tarasenko et al. 2010).

An excess or deficiency of membrane cholesterol appears to cause its redistribution between the plasma and vesicular membranes, thereby altering the ability of vesicles to accumulate the protons and neurotransmitters. The application of MβCD complexed with cholesterol to the synaptosomes led to an increase in proton accumulation within synaptic vesicles (in contrast, the addition of MβCD alone caused dissipation of the proton gradient). This result is in a good agreement with the data of Yoshinaka et al. (2004) and demonstrates that isolated synaptic vesicles (similarly with the synaptosomes) are able to lose or accumulate protons depending on whether MβCD is applied alone or in "saturated" complex with cholesterol. In this case, MβCD-cholesterol complex delivers cholesterol to the plasma membrane (Christian et al. 1997; Zamir and Charlton 2006), and thereafter extra cholesterol appears to be transferred to synaptic vesicle membranes (Tarasenko et al. 2010).

Taken together, our data support the evidence that the proper level of membrane cholesterol is of fundamental importance for efficient neurotransmission. Acute depletion or replenishment of plasma membrane cholesterol causes the changes not only in the plasma membrane structures, but influences the functional state of synaptic vesicles, thereby altering the processes of release of neurotransmitters.

References

Allen NJ, Attwell D (2004) The effect of simulated ischaemia on spontaneous GABA release in area CA1 of the juvenile rat hippocampus. J Physiol 561:485–498

Allen NJ, Rossi DJ, Attwell D (2004) Sequential release of GABA by exocytosis and reversed uptake leads to neuronal swelling in simulated ischemia of hippocampal slices. J Neurosci 24:3837–3849

Baker DA, Xi Z, Shen H, Swanson CJ, Kalivas PW (2002) The origin and neuronal function of in vivo nonsynaptic glutamate. J Neurosci 22:9134–9141

Barnes K, Ingram JC, Bennett MDM et al (2004) Methyl-beta-cyclodextrin stimulates glucose uptake in Clone 9 cells: a possible role for lipid rafts. Biochem J 378:343–351

Borisova T, Krisanova N, Sivko R, Borysov A (2010a) Cholesterol depletion attenuates tonic release but increases the ambient level of glutamate in rat brain synaptosomes. Neurochem Int 56:466–478

Borisova T, Sivko R, Borysov A, Krisanova N (2010b) Diverse presynaptic mechanisms underlying methyl-beta-cyclodextrin—mediated changes in glutamate transport. Cell Mol Neurobiol 30:1013–1023

Cavelier P, Attwell D (2005) Tonic release of glutamate by a DIDS-sensitive mechanism in rat hippocampal slices. J Physiol 564:397–410

Cidon S, Sihra T (1989) Characterization of an H^+-ATPase in rat brain synaptic vesicles. J Biol Chem 264:8281–8288

Christian AE, Haynes MP, Phillips MC, Rothblat GH (1997) Use of cyclodextrins for manipulating cellular cholesterol content. J Lipid Res 38:2264–2272

Deutsch JW, Kelly RB (1981) Lipids of synaptic vesicles: relevance to the mechanism of membrane fusion. Biochemistry 20:378–385

Gil C, Soler-Jover A, Blasi J (2005) Aguilera synaptic proteins and SNARE complexes are localized in lipid rafts from brain synaptosomes. Biochem Biophys Res Commun 329:117–124

Holz RW, Senter RA (1985) Plasma membrane and chromaffin granule characteristics in digitonin-treated chromaffin cells. J Neurochem 45:1548–1557

Huss M, Wieczorek H (2009) Inhibitors of V-ATPases: old and new players. J Exp Biol 212:341–346

Jabaudon D, Shimamoto K, Yasuda-Kamatani Y (1999) Inhibition of uptake unmasks rapid extracellular turnover of glutamate of nonvesicular origin. Proc Natl Acad Sci USA 96:8733–8738

Jadot M, Andrianaivo F, Dubois F, Wattiaux R (2001) Effects of methylcyclodextrin on lysosomes. Eur J Biochem 268:1392–1399

Katz J, Wals PA (1985) Studies with digitonin-treated rat hepatocytes (nude cells). J Cell Biochem 28:207–228

Krisanova N, Sivko R, Kasatkina L, Borisova T (2012) Neuroprotection by lowering cholesterol: a decrease in membrane cholesterol content reduces transporter-mediated glutamate release from brain nerve terminals. Biochim Biophys Acta 1822:1013–1023

Krisanova NV, Ttiksh IO, Borisova TA (2009) Synaptopathy under conditions of altered gravity: changes in synaptic vesicle fusion and glutamate release. Neurochem Int 55:724–731

Lang T, Bruns D, Wenzel D et al (2001) SNAREs are concentrated in cholesterol-dependent clusters that define docking and fusion sites for exocytosis. EMBO J 20:2202–2213

Lange Y, Ye J, Steck TL (2004) How cholesterol homeostasis is regulated by plasma membrane cholesterol in excess of phospholipids. Proc Natl Acad Sci USA 101:11664–11667

Maltseva EL, Palmina NP, Pryme IF (1991) The effect of a phorbol ester on the lipid microviscosity of two endoplasmic reticulum membrane fractions isolated from Krebs II ascites cells. J Cell Biochem 46:260–265

Maltseva EL, Borovok NV, Zlatanov I (1996) The relation between the changes in viscosity, lipid peroxidation indices and adenylate cyclase activity in plasma membranes of liver and brain cells after injection of antioxidant. Membr Cell Biol 9:621–630

Perez-Castiñeira JR, Apps DK (1990) Vacuolar H(+)-ATPase of adrenal secretory granules. Rapid partial purification and reconstitution into proteoliposomes. Biochem J 271:127–131

Rossi DJ, Hamann M, Attwell D (2003) Multiple modes of GABAergic inhibition of rat cerebellar granule cells. J Physiol 548:97–110

Rutledge EM, Aschner M, Kimelberg HK (1998) Pharmacological characterization of swelling-induced D-[3H]aspartate release from primary astrocyte cultures. Am J Physiol 274:1511–1520

Salaun C, James DJ, Chamberlain LH (2004) Lipid rafts and the regulation of exocytosis. Traffic 5:1–10

Sivko R, Krisanova N, Borisova T (2012) Leakage of L-[14C]glutamate from cholesterol-deficient rat brain nerve terminals at low temperature conditions. Neurobiol Lipids 10:1–6. http://neuro-biologyoflipids.org/content/10/1/

Steck TL, Ye J, Lange Y (2002) Probing red cell membrane cholesterol movement with cyclopdextrin. Biophys J 83:2118–2125

Tarasenko AS, Sivko RV, Krisanova NV, Himmelreich NH, Borisova TA (2010) Cholesterol depletion from the plasma membrane impairs proton and glutamate storage in synaptic vesicles of nerve terminals. J Mol Neurosci 41:358–367

Taverna E, Saba E, Rowe J, Francolini M, Clementi F, Rosa P (2004) Role of lipid microdomains in P/Q-type calcium channel ($Ca_v2.1$) clustering and function in presynaptic membranes. J Biol Chem 279:5127–5134

Thiele C, Hannah MJ, Fahrenholz F, Huttner WB (2000) Cholesterol binds to synaptophysin and is required for biogenesis of synaptic vesicles. Nat Cell Biol 2:42–49

Tretter L, Chinopoulos C, Adam-Vizi V (1998) Plasma membrane depolarization and disturbed Na+ homeostasis induced by the protonophore carbonyl cyanide-p-trifluoromethoxyphenyl-hydrazon in isolated nerve terminals. Mol Pharm 53:734–774

Wasser CR, Ertünc M, Liu X et al (2007) Cholesterol-dependent balance between evoked and spontaneous vesicle recycling. J Physiol 579(2):413–429

Xia F, Gao X, Kwan E et al (2004) Disruption of pancreatic β-cells lipid rafts modifies Kv2.1 channel gating and insulin exocyrtosis. J Biol Chem 279:24685–24691

Xia F, Leung YM, Gaisano G et al (2007) Targeting of Kv4, Cav1.2 and SNARE proteins to cholesterol-rich lipid rafts in pancreatic a-cells: effects on glucagons stimulus-secretion coupling. Endocrinology 148:2157–2167

Yancey PG, Rodrigueza WV, Kilsdonk EP et al (1996) Cellular cholesterol efflux mediated by cyclodextrins. J Biol Chem 271:16026–16034

Yoshinaka K, Kumanogoh H, Nakamura S, Maekawa S (2004) Identification of V-ATPase as a major component in the raft fraction prepared from the synaptic plasma membrane and the synaptic vesicle of rat brain. Neurosci Lett 363:168–172

Zoccarato F, Cavallini L, Alexandre A (1999) The pH-sensitive dye acridine orange as a tool to monitor exocytosis/endocytosis in synaptosomes. J Neurochem 72:625–633

Zhang J, Feng Y, Forgac M (1994) Proton conduction and bafilomycin binding by the V0 domain of the coated vesicle V-ATPase. J Biol Chem 269:23518–23523

Zhou Q, Petersen CC, Nicoll RA (2000) Effects of reduced vesicular filling on synaptic transmission in rat hippocampal neurons. J Physiol 525:195–206

Zamir O, Charlton MP (2006) Cholesterol and synaptic transmitter release at crayfish neuromuscular junctions. J Physiol 571:83–99

Chapter 6
Neuroprotection by Lowering Cholesterol

Abstract In stroke, cerebral hypoxia/ischemia, and traumatic brain injury, the development of neurotoxocity is provoked by enhanced extracellular glutamate, which is released from nerve cells mainly by glutamate transporter reversal—a distinctive feature of these pathological states. Transporter-mediated glutamate release from the synaptosomes: (1) stimulated by depolarization of the plasma membrane; (2) by means of heteroexchange with competitive transportable inhibitor of glutamate transporters DL-threo-β-hydroxyaspartate; (3) in low-[Na^+] medium; and (4) during dissipation of the proton gradient of synaptic vesicles by the protonophore FCCP; was decreased under conditions of cholesterol deficiency by approximately 24, 28, 40, 17 %, respectively.

A decrease in the level of membrane cholesterol attenuated transporter-mediated glutamate release from nerve terminals. Therefore, lowering cholesterol may be used in neuroprotection in stroke, ischemia, and traumatic brain injury that are associated with an increase in glutamate uptake reversal. This data can explain the neuroprotective effects of statins in these pathological states and provide one of the mechanisms of their neuroprotective action. However, besides these disorders lowering cholesterol may cause harmful consequences decreasing glutamate uptake by nerve terminals.

In stroke, cerebral hypoxia/ischemia, traumatic brain injury, and energy deprivation, glutamate is released from the nerve terminals into the extracellular space via glutamate transporters, thereby causing neurotoxicity, whereas besides these pathological states transporters predominantly operate in the inward direction (Fig. 6.1a). Therefore, transporter-mediated glutamate release from the neurons mainly contributes to an increase in ambient glutamate concentration under pathological conditions. An increase in transporter-mediated release, and thus the extracellular level of the neurotransmitter may alter the functioning of NMDA receptors, which contribute to cognitive capacities and pathology (Xu et al. 2012). Since the activity of glutamate

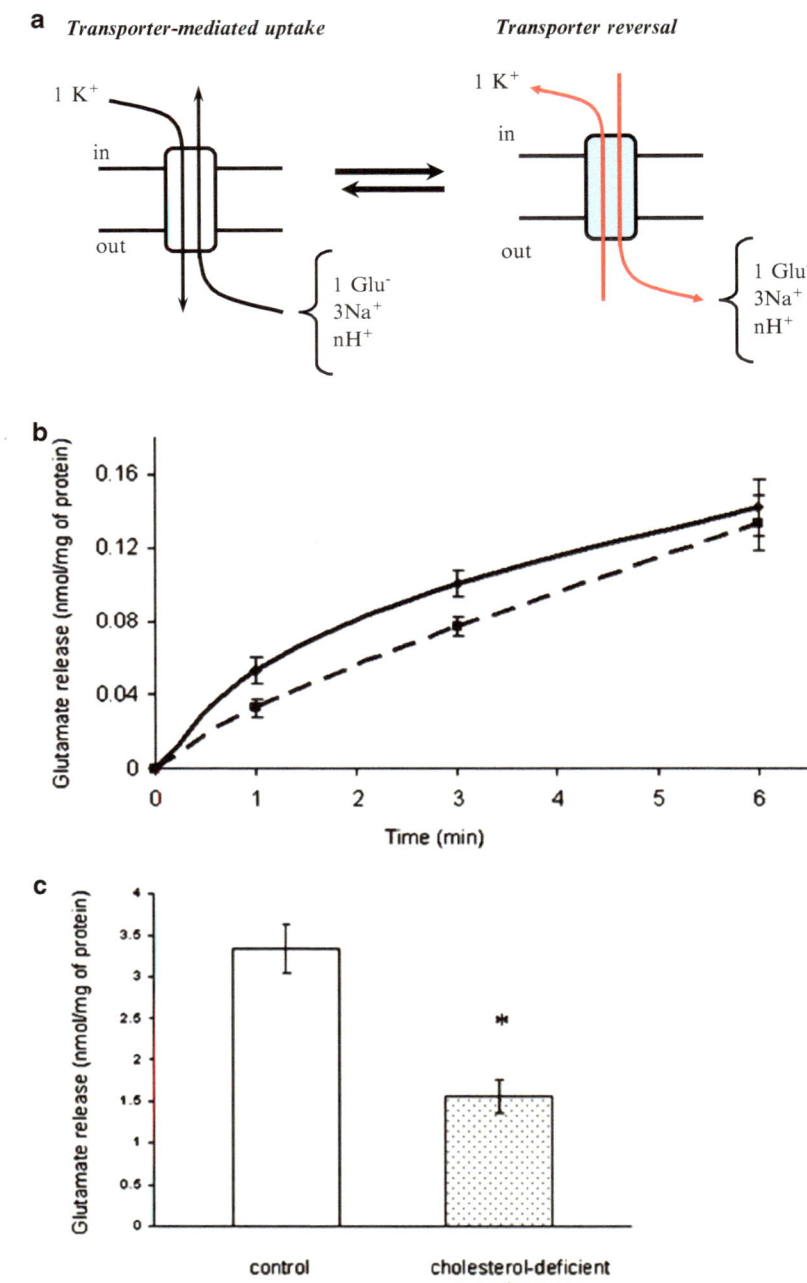

Fig. 6.1 (**a**) Glutamate transporters acting in the direct and reverse modes. Stimulated by high-KCl (35 mM) transporter-mediated release of preloaded L-[^{14}C]glutamate (**b**) and cold glutamate (glutamate dehydrogenase assay) (**c**) from control synaptosomes (*solid line* in (**b**); *empty column*

transporters in the inward direction is decreased in cholesterol deficiency, we hypothesized that their reverse function could also be attenuated. It is clear that a delay in elevation of ambient glutamate has a potential for preventing brain damage under these pathological states.

6.1 Stimulated by Depolarization of the Plasma Membrane Transporter-Mediated Release of Glutamate from Cholesterol-Deficient Nerve Terminals

In the experiments, depolarization of the plasma membrane of the nerve terminals by high-KCl in Ca^{2+}-free medium causes reversal of glutamate transporters and release of glutamate from the cytosol (Fig. 6.1a). The value of transporter-mediated release of L-[^{14}C]glutamate from the synaptosomes measured at 3 min time point was decreased by ~24 % after cholesterol extraction with 15 mM MβCD and washing of the acceptor from the media (LP) that was equal to 0.101325 ± 0.006750 nmol/mg of protein in the control and 0.0772 ± 0.0048 nmol/mg of protein in cholesterol deficiency ($P \leq 0.05$, Student's t-test, $n = 5$) (Fig. 6.1b). (However, at 6 min time point it was equal to 0.141855 ± 0.008865 nmol/mg of protein in the control and 0.136065 ± 0.008865 nmol/mg of protein in MβCD-treated synaptosomes). Similar to the previous chapters, the application of 15 mM MβCD complexed with cholesterol (2.3 mM) did not induce changes in transporter-mediated release of L-[^{14}C] glutamate from the synaptosomes as compared to the control. Thus, data on high-KCl-evoked transporter-mediated release of L-[^{14}C]glutamate from the synaptosomes showed that the initial velocity of this release was decreased under conditions of cholesterol deficiency (Krisanova et al. 2012).

To measure net transporter-mediated release of glutamate from cholesterol-depleted synaptosomes, it was plausible to inhibit glutamate uptake in order to neglect its contribution. It is so because released glutamate is continuously removed from the extracellular medium by glutamate transporters with different effectiveness in the control and cholesterol deficiency. In this case, the application of glutamate transporter inhibitors is not possible because non-transportable inhibitor DL-TBOA attenuates as direct as reversed transport of glutamate, whereas transportable inhibitor DL-threo-β-hydroxyaspartate (DL-THA) causes release of glutamate by means of heteroexchange. Weak uptake shown in cholesterol-depleted

---

Fig. 6.1 (continued) in (**c**)) and the synaptosomes preliminary treated with 15 mM MβCD (*dash line* in (**b**); *dotted column* in (**c**)). Control and MβCD-treated synaptosomes were loaded with L-[^{14}C]glutamic acid (1 nmol/mg of protein, 238 mCi/mmol) (**b**) or 50 μM cold glutamate (**c**). After loading, samples (0.5 mg of protein/mL) were preincubated for 8 min at 37 °C, then at different time points the aliquots of the samples were centrifuged and L-[^{14}C]glutamate radioactivity was determined (**b**) or fluorescence intensity of NADH was measured at 6 min time point using glutamate dehydrogenase assay (**c**). Data are means ± SEM of five independent experiments, each performed in triplicate. *$P \leq 0.05$ as compared to the control. Figure as in Krisanova et al. (2012)

nerve terminals should increase the extracellular glutamate concentration, and thus apparent release of L-[^{14}C]glutamate. Therefore, it was reasonable to analyze depolarization-evoked transporter-mediated release of endogenous glutamate from cholesterol-deficient synaptosomes using glutamate dehydrogenase assay. This release could not be registered in the synaptosomes after the treatment with MβCD (however, the rest of tonic release was detected under these conditions, see the previous chapters). It is so because the treatment with MβCD for 30 min followed by washing procedure with 10 volumes of standard solution led to the removal of the large amount of endogenous glutamate from the synaptosomes. This data is in accordance with the study that the addition of 15 mM MβCD caused dissipation of the proton gradient of synaptic vesicles and massive release of L-[^{14}C]glutamate from the synaptosomes (Fig. 5.1). In the experiments, cold (nonradioactive) glutamate was preloaded to MβCD-treated synaptosomes and depolarization-evoked transporter-mediated release of the neurotransmitter was measured using glutamate dehydrogenase assay. As shown in Fig. 6.1c, stimulated by high-KCl synaptosomal glutamate release in Ca^{2+}-free medium for 6 min was decreased after cholesterol depletion and consisted of 3.3333 ± 0.3400 nmol/mg of protein in the control and 1.5606 ± 0.1600 nmol/mg of protein in cholesterol deficiency. This data was in accordance with the above results obtained with radioactive L-[^{14}C]glutamate (Fig. 6.1b), where a significant decrease in the velocity of transporter-mediated L-[^{14}C]glutamate release at 1 and 3 min time points was demonstrated under conditions of cholesterol deficiency. However, the difference in this release between the control and cholesterol-deficient synaptosomes presented in Fig. 6.1b (6 min point) was not so significant as in Fig. 6.1c. It is so because of the existence of L-[^{14}C] glutamate uptake, which, in addition, is more effective in the control than in MβCD-treated synaptosomes. In contrast, glutamate dehydrogenase experiments showed "pure" transporter-mediated release, when released glutamate was metabolized by the enzyme (Krisanova et al. 2012).

6.2 Release of Glutamate from Cholesterol-Deficient Nerve Terminals by Means of Heteroexchange

Heteroexchange and transporter-mediated release of glutamate have a common rate-limiting step in the transport process, and so heteroexchange may be used for the evaluation of transporter-mediated release of glutamate (Jabaudon et al. 2000). Release of L-[^{14}C]glutamate by means of heteroexchange was evaluated with transportable inhibitor of glutamate transporters DL-THA, which is a substrate for glutamate transporters, competitively inhibits glutamate uptake, but does not prevent molecular transport mechanism.

Release of L-[^{14}C]glutamate by heteroexchange with 100 μM DL-THA was decreased (by ~28 %) at 6 min time point from 0.137995 ± 0.009650 nmol/mg of protein to 0.099395 ± 0.008680 nmol/mg of protein as a result of the treatment of the nerve terminals with 15 mM MβCD ($P \leq 0.05$, Student's t-test, $n = 4$) (Fig. 6.2).

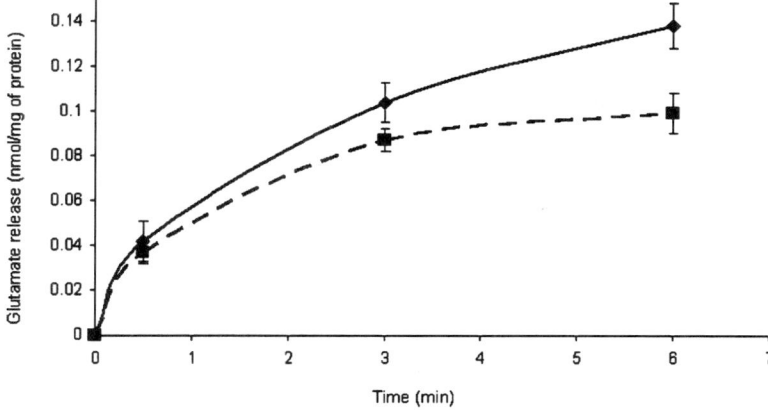

Fig. 6.2 Release of L-[¹⁴C]glutamate by means of heteroexchange stimulated by transportable inhibitor of glutamate transporters DL-THA (100 µM) from control synaptosomes (*solid line*) and cholesterol-deficient synaptosomes (*dash line*). After loading of control and MβCD-treated synaptosomes with L-[¹⁴C]glutamate, samples (0.5 mg of protein/mL) were preincubated for 8 min at 37 °C, at 1, 3, and 6 min time points, the aliquots of the samples were centrifuged. Data are means ± SEM of four independent experiments, each performed in triplicate. Figure as in Krisanova et al. (2012)

Using MβCD complexed with cholesterol (2.3 mM cholesterol in 15 mM MβCD), it was demonstrated that the treatment with the complex did not change significantly heteroexchange of L-[¹⁴C]glutamate with DL-THA in comparison with the control. Taking into account that release of glutamate by heteroexchange is justified for the evaluation of transporter-mediated glutamate release, it is clear that the above results confirmed a decrease in glutamate transporter reversal under conditions of cholesterol deficiency. This result is in accord with the above data on stimulated by depolarization transporter-mediated release of glutamate from cholesterol-depleted synaptosomes (Krisanova et al. 2012).

6.3   Release of Glutamate in Low-Na⁺ Medium from Cholesterol-Deficient Nerve Terminals

A driving force for glutamate uptake by transporters is Na⁺/K⁺ gradient of the plasma membrane of the nerve cells. Thus, a reduction in the extracellular Na⁺ concentration (from 126 mM up to 21 mM) is expected to inhibit uptake and facilitate the reversal of transporters resulting in release of cytoplasmic glutamate into the extracellular space. Using monovalent organic cations *N*-methyl-D-glucamine (NMDG) to replace extracellular Na⁺, it was found that L-[¹⁴C]glutamate release for 6 min was equal to 0.066585 ± 0.005790 nmol/mg of protein in the control and 0.039565 ± 0.005790 nmol/mg of protein in cholesterol-depleted synaptosomes

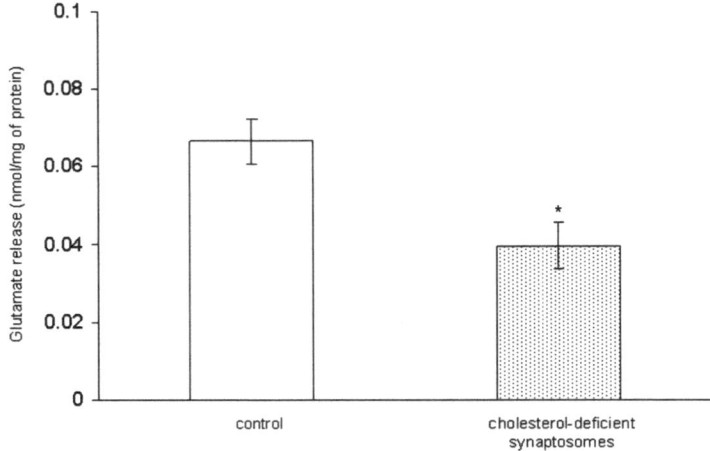

Fig. 6.3 Release of L-[^{14}C]glutamate from control (*empty column*) and cholesterol-depleted (*dotted column*) synaptosomes in low-Na$^+$ (21 mM Na$^+$, 105 mM NMDG), Ca^{2+}-free incubation media for 6 min. After loading of control and MβCD-treated synaptosomes with L-[^{14}C]glutamate, samples (0.5 mg of protein/mL) were preincubated for 8 min at 37 °C, then incubated for 6 min, rapidly sedimented in a microcentrifuge. Data are means ± S.E.M. of four independent experiments, each performed in triplicate. *$P \leq 0.05$ as compared to the control. Figure as in Krisanova et al. (2012)

($P \leq 0.05$, Student's *t*-test, $n = 4$) (Fig. 6.3). In the nerve terminals treated with 15 mM MβCD complexed with 2.3 mM cholesterol, changes in L-[^{14}C]glutamate release in low-Na$^+$ medium were not found as compared to the untreated control. Thus, it was demonstrated that the value of L-[^{14}C]glutamate release from cholesterol-deficient nerve terminals in low-Na$^+$ medium was lesser than that from the control synaptosomes (Krisanova et al. 2012).

6.4 Transporter-Mediated Release of Glutamate from Cholesterol-Deficient Nerve Terminals During Dissipation of the Proton Gradient of Synaptic Vesicles

During application of the protonophore FCCP, which dissipates the proton gradient and inhibits uptake of glutamate by synaptic vesicles, L-[^{14}C]glutamate release from control and cholesterol-depleted synaptosomes was assessed and it was revealed that this release became lesser in cholesterol deficiency as compared to the control. As shown in Fig. 6.4 (the first pair of columns), FCCP-evoked release of L-[^{14}C] glutamate for 6 min was equal to 0.101711 ± 0.004300 nmol/mg of protein in the control and 0.084534 ± 0.004300 nmol/mg of protein in cholesterol-deficient synaptosomes ($P \leq 0.05$, Student's *t*-test, $n = 4$). High-KCl was applied at 5 min time point after the addition of FCCP to expand further synaptosomal transporter-mediated

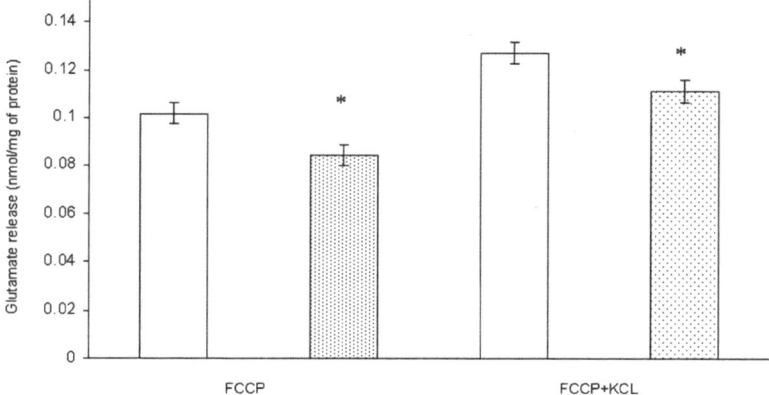

Fig. 6.4 Release of L-[^{14}C]glutamate in the presence of the protonophore FCCP (1 μM) and FCCP + KCl (1 μM, 35 mM, respectively) from control (*empty columns*) and cholesterol-deficient (*dotted columns*) nerve terminals. After loading of control and MβCD-treated synaptosomes with L-[^{14}C]glutamate, samples (0.5 mg of protein/mL) were preincubated for 8 min at 37 °C, then incubated for 6 min and rapidly sedimented in a microcentrifuge (FCCP-experiments). In FCCP + KCl experiments, KCl was added at 5 min time point after the addition of FCCP, then the synaptosomes were incubated for 6 min and rapidly sedimented. Data are means ± S.E.M. of four independent experiments, each performed in triplicate. *$P \leq 0.05$ as compared to the control. Figure as in Krisanova et al. (2012)

release of L-[^{14}C]glutamate. It was found that cholesterol deficiency decreased the latest from 0.127187 ± 0.004300 nmol/mg of protein in the control to 0.111458 ± 0.004300 nmol/mg of protein in MβCD-treated synaptosomes ($P \leq 0.05$, Student's *t*-test, $n = 4$) (Fig. 6.4, the second pair of columns) (Krisanova et al. 2012).

6.5   Transporter-Mediated Release of Glutamate from Cholesterol-Deficient Nerve Terminals Under Conditions of Energy Deprivation

In energy deprivation experiments, iodoacetate (a potent inhibitor of G3P dehydrogenase) and rotenone (inhibitor of mitochondrial Complex I)/oligomycin (inhibitor of ATP synthase) were used. High-KCl was added at 7 min time point after the addition of iodoacetate or rotenone/oligomycin to expand transporter-mediated release of L-[^{14}C]glutamate, and then the initial velocity of this process was measured. As shown in Fig. 6.5 (the first and second columns), after the application of iodoacetate (1 mM) and KCl (35 mM) to the synaptosomes transporter-mediated release of L-[^{14}C]glutamate for 10 min was decreased from 0.14668 ± 0.01000 nmol/mg of protein in the control to 0.0801 ± 0.00800 nmol/mg of protein in cholesterol-deficient synaptosomes (15 mM MβCD) ($P \leq 0.05$, Student's *t*-test, $n = 3$). The treatment of the synaptosomes with rotenone/oligomycin (4 μM/4 μg/mL, respectively)

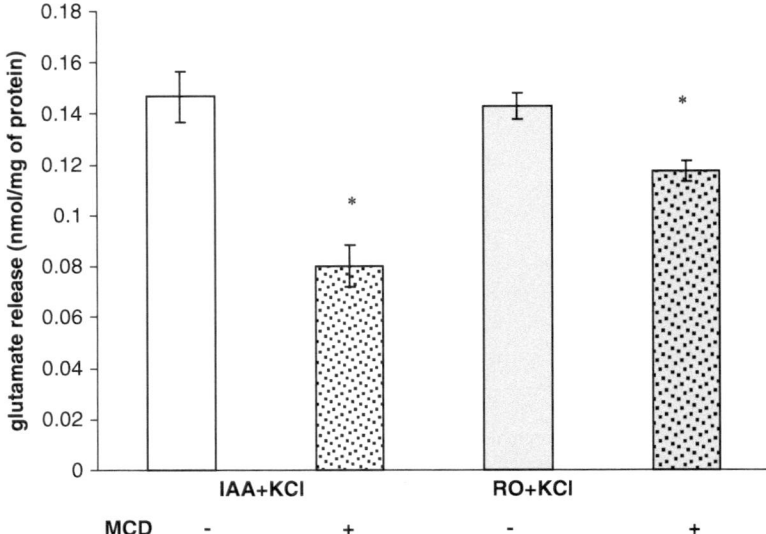

Fig. 6.5 Release of L-[¹⁴C]glutamate in the presence of iodoacetate and KCl (1 mM and 35 mM, respectively) (the first and second columns) and rotenone/oligomycin and KCl (4 μM/4 μg/mL and 35 mM, respectively) (the third and fourth columns) from control (*clear columns*) and cholesterol-deficient (*dotted columns*) nerve terminals. After loading of control and MβCD-treated synaptosomes with L-[¹⁴C]glutamate, samples (0.5 mg of protein/mL) were preincubated for 7 min at 37 °C with iodoacetate or rotenone/oligomycin, then after the addition of KCl were incubated for 3 min and rapidly sedimented in a microcentrifuge. Data are means ± S.E.M. of three independent experiments, each performed in triplicate. *$P \leq 0.05$ as compared to the control. Figure as in Krisanova et al. (2012)

and KCl (35 mM) resulted in a reduction of transporter-mediated release of L-[¹⁴C] glutamate from 0.14282 ± 0.00500 nmol/mg of protein in the control to 0.11725 ± 0.00400 nmol/mg of protein in cholesterol-deficient synaptosomes ($P \leq 0.05$, Student's t-test, $n = 4$) (Fig. 6.5, the third and fourth columns) (Krisanova et al. 2012).

Thus, using cholesterol-deficient rat brain nerve terminals, radiolabeled technique and glutamate dehydrogenase assay, it was shown a decrease in transporter-mediated release of glutamate: (1) stimulated by the depolarization of the plasma membrane; (2) by means of heteroexchange with DL-THA; (3) in low-Na⁺ medium; (4) during dissipation of the proton gradient of synaptic vesicles; (5) under conditions of energy deprivation. As transporter-mediated release of glutamate from nerve cells provokes the development of exitotoxicity in stroke, cerebral hypoxia/ischemia, and traumatic brain injury, the lowering of the level of membrane cholesterol, which attenuates the velocity of release, can have neuroprotective effect in these pathological states. It is clear that the lesser transporter-mediated glutamate release is, the slower the development of neurotoxicity. Abulrob et al. (2005) showed the neuroprotective activity of MβCD against oxygen-glucose deprivation in cortical neuronal cultures that could support above suggestion.

MβCD is used for the extraction of cholesterol in the experiments, so the question rose: "Can it be employed for neuroprotection by direct application to injured brain regions?" Recently, it was shown that the addition of MβCD to the nerve terminals and blood platelets caused dissipation of the proton gradient of their acidic compartments, i.e., synaptic vesicles and secretory granules, respectively (Borisova et al. 2010b, 2011). This dissipation was accompanied with a dramatic increase in glutamate release from the nerve terminals, thus the presence of MβCD per se may cause an elevation of the extracellular glutamate concentration. Intravenous administration of MβCD seems to be not effective because of its low permeability to brain–blood barrier (Sierra et al. 2011; Camargo et al. 2001; Monnaert et al. 2004) and its possible harmful influence on blood components. It was shown that MβCD decreasing the level of cholesterol in blood platelets caused a reduction of glutamate uptake (Borisova et al. 2011). As consequence, it can result in an increase in the glutamate concentration in the plasma, and then in the cerebrospinal fluid (because of glutamate balance between them (O'Kane et al. 1999)), thereby changing extracellular glutamate homeostasis in the brain (Krisanova et al. 2012).

In the study on the assessment of glutamate transporter reversal at low level of membrane cholesterol, one of the main uncertainties is the fact that in the experiments with radiolabeled glutamate, a decrease in transporter-mediated glutamate release from cholesterol-deficient nerve terminals has been registered against a background of enhanced extracellular glutamate concentration. As transporter-mediated release of glutamate depends from the ratio glu_{ex}/glu_{int}, a decrease in the ratio per se may attenuate this release because of thermodynamic reasons. In addition, cholesterol deficiency reduces the initial velocity of glutamate uptake (Butchbach et al. 2004; Borisova et al. 2010a) that may bring inaccuracy to the measurements. However, in the pathological states accompanied by glutamate uptake reversal, glutamate transporter activity in the inward direction can be negligible (Krisanova et al. 2012).

It can be suggested that the attenuation of transporter-mediated glutamate release from the nerve terminals in cholesterol deficiency is a result of the reduction in the activity of glutamate transporters per se in response to the changes in the lipid surroundings. The data of Butchbach et al. (2004) showed that a loss of cholesterol led to the disturbance in cluster organization of glutamate transporters and changes in their trafficking in primary cortical cultures. Moreover, up- and down-regulation of glutamate transporter activity by exocytosis-like trafficking of transporter-containing vesicles (Robinson 2006) may be affected because of the changes of the exocytotic process in cholesterol deficiency (Lang et al. 2001; Jennings et al. 1999; Borisova et al. 2010b).

This data can explain neuroprotective features of statins, widely applicable cholesterol-reducing drugs, which are selective inhibitors of 3-hydroxyl-3-methylglutaryl coenzyme A reductase, the rate-limiting enzyme of the mevalonate pathway for cholesterol biosynthesis. The data of the literature showed that the treatment with statins diminished the level of cholesterol in the brain (Kirsch et al. 2003). Sierra et al. (2011) found that monacolin J derivatives (natural and semisynthetic statins) were the best candidates for the prevention of neurodegeneration due to their high capacity for penetration of the brain–blood barrier and cholesterol lowering effect on neurons. Increasing evidence indicates that statins, e.g., simvastatin, atorvastatin, and pravastatin, may be beneficial during acute stroke as well as post-ischemically in animal models and have neuroprotective features under conditions of cerebral ischemia, traumatic brain injury, and excitotoxic amino acid exposure (Nagaraja et al. 2006; Sironi et al. 2003; Berger et al. 2008; Chen et al. 2005; Zhang et al. 2005; Sugiura et al. 2007). The data of Funck et al. (2011) demonstrated that atorvastatin treatment had effects on pentylenetetrazol-induced seizures. Berger et al. (2008) showed that high dose of pravastatin administered repetitively after stroke onset improved neurological outcome. After induction of focal ischemia in rats, it was found that the treatment with statins reduced the extent of brain damage (Sironi et al. 2003). Therefore, the data of the literature suggests that statins activate neuroprotective mechanisms; however, its nature is far from being clear (Krisanova et al. 2012).

Several hypotheses are forwarded on the possible mechanisms of the neuroprotective effect of statins. The first ones consider that the neuroprotective action of statins is independent of cholesterol reduction. There are evidences that statins act on the nitric oxide synthase system (Endres et al. 1998) and inhibit release of potentially damaging cytokines such as IL-6 in the early phase of cerebral ischemia (Berger et al. 2008). It may involve non-sterol mechanisms based on the effects on the endothelial cells, macrophages, platelets, and smooth muscle cells (Berger et al. 2008; Hess et al. 2000). The second point of view is that statins and MβCDs protect neurons from the death changing excessive stimulation of NMDA receptors (Bösel et al. 2005; Xu et al. 2012; Zacco et al. 2003; Wang et al. 2009). It was demonstrated that a decrease in cholesterol level by simvastatin in primary neuronal cultures protected the cells from NMDA-induced neuronal damage probably by reducing the association of NMDA receptors to lipid rafts (Ponce et al. 2008). Abulrob et al. (2005) also suggested that cholesterol extraction from detergent-resistant microdomains affected NMDA receptor subunit distribution and signal propagation resulting in neuroprotection of cortical neuronal cultures against ischemic and excitotoxic insults. Modulation of NMDA receptors after simvastatin treatment could explain their anxiolytic-like activity and anti-inflammatory mechanisms in the experimental model of Parkinson's disease (Yan et al. 2011). It was revealed that simvastatin reduced the deleterious effects caused by kainate, including the severity of seizures, excitotoxicity, and oxidative damage in the hippocampus and other limbic structures of the brain cortex (Ramirez et al. 2011).

The possible mechanism of neuroprotective action of statins was proposed based on the data that the reduction of the level of membrane cholesterol decreased transporter-mediated glutamate release from the nerve terminals. This suggestion does not contradict NMDA-dependent and non-sterol mechanisms of statin action (even may have additive or synergetic effect), but seems to be actual for the early phase of neuroprotection. The data of Berger et al. (2008) may be considered in support of this hypothesis over the others. Using cerebral microdialyis in a temporary middle cerebral artery occlusion model in Wistar rats, the authors demonstrated that an increase in the extracellular level of striatal glutamate in the ischemic hemisphere was attenuated by pravastatin compared to placebo (Berger et al. 2008).

It may be supposed that a decrease in cholesterol content in these experiments most likely provoked a reduction of transporter-mediated release of glutamate thereby causing a reduction in the extracellular glutamate concentration in the ischemic hemisphere (Krisanova et al. 2012).

So, it was shown that a decrease in the level of membrane cholesterol in the nerve terminals reduced transporter-mediated glutamate release. The latest is the main mechanism underlying the enhancement of the concentration of neurotoxic extracellular glutamate in stroke, cerebral hypoxia/ischemia, hypoglicemia, and traumatic brain injury. Therefore, a reduction of cholesterol content may be used for neuroprotection under these pathological conditions, i.e., "neuroprotection by lowering cholesterol." This data may explain the neuroprotective effect followed by the administration of statins in stroke, cerebral hypoxia/ischemia, seizures, excitotoxicity, oxidative damage, and traumatic brain injury. However, besides these pathologies, the normal level of membrane cholesterol is very important for proper synaptic transmission and a decrease in membrane cholesterol content of the nerve terminals may cause neurotoxic consequences because of weak glutamate uptake and the enlargement of the extracellular glutamate concentration (Krisanova et al. 2012).

References

Abulrob A, Tauskela JS, Mealing G, Brunette E, Faid K, Stanimirovic D (2005) Protection by cholesterol-extracting cyclodextrins: a role for N-methyl-D-aspartate receptor redistribution. J Neurochem 92:1477–1486

Berger C, Xia F, Maurer MH, Schwab S (2008) Neuroprotection by pravastatin in acute ischemic stroke in rats. Brain Res Rev 58:48–56

Borisova T, Krisanova N, Sivko R, Borysov A (2010a) Cholesterol depletion attenuates tonic release but increases the ambient level of glutamate in rat brain synaptosomes. Neurochem Int 56:466–478

Borisova T, Sivko R, Borysov A, Krisanova N (2010b) Diverse presynaptic mechanisms underlying methyl-beta-cyclodextrin—mediated changes in glutamate transport. Cell Mol Neurobiol 30:1013–1023

Borisova T, Kasatkina L, Ostapchenko L (2011) The proton gradient of secretory granules and glutamate transport in blood platelets during cholesterol depletion of the plasma membrane by methyl-beta-cyclodextrin. Neurochem Int 59:965–975

Bösel J, Gandor F, Harms C et al (2005) Neuroprotective effects of atorvastatin against glutamate-induced excitotoxicity in primary cortical neurones. J Neurochem 92:1386–1398

Butchbach M, Tian G, Guo H, Lin CL (2004) Association of excitatory amino acid transporters, especially EAAT2, with cholesterol-rich lipid raft microdomains. J Biol Chem 279:34388–34396

Camargo F, Erickson RP, Garver WS et al (2001) Cyclodextrins in the treatment of a mouse model of Niemann-Pick C disease. Life Sci 70:131–142

Chen J, Zhang C, Jiang H et al (2005) Atorvastatin induction of VEGF and BDNF promotes brain plasticity after stroke in mice. J Cereb Blood Flow Metab 25:281–290

Endres M, Laufs U, Huang Z, Nakamura T, Huang P, Moskowitz MA, Liao JK (1998) Stroke protection by 3-hydroxy-3-methylglutaryl (HMG)-CoA reductase inhibitors mediated by endothelial nitric oxide synthase. Proc Natl Acad Sci USA 95:8880–8885

Funck VR, Oliveira CV, Pereira LM et al (2011) Differential effects of atorvastatin treatment and withdrawal on pentylenetetrazol-induced seizures. Epilepsia 52:2094–2104

Hess DC, Demchuk AM, Brass LM, Yatsu FM (2000) HMG-CoA reductase inhibitors (statins): a promising approach to stroke prevention. Neurology 54:790–796

Jabaudon D, Scanziani M, Gähwiler BH, Gerber U (2000) Acute decrease in net glutamate uptake during energy deprivation. Proc Natl Acad Sci USA 97:5610–5615

Jennings LJ, Xu QW, Firth TA (1999) Cholesterol inhibits spontaneous action potentials and calcium currents in guinea pig gallbladder smooth muscle. Am J Physiol 277:1017–1026

Kirsch C, Eckert GP, Mueller WE (2003) Statin effects on cholesterol micro-domains in brain plasma membranes. Biochem Pharmacol 65:843–856

Krisanova N, Sivko R, Kasatkina L, Borisova T (2012) Neuroprotection by lowering cholesterol: a decrease in membrane cholesterol content reduces transporter-mediated glutamate release from brain nerve terminals. Biochim Biophys Acta 1822:1013–1023

Lang T, Bruns D, Wenzel D, Riedel D, Holroyd P, Thiele C, Jahn R (2001) SNAREs are concentrated in cholesterol-dependent clusters that define docking and fusion sites for exocytosis. EMBO J 20:2202–2213

Monnaert V, Tilloy S, Bricout H, Fenart L, Cecchelli R, Monflier E (2004) Behavior of alpha-, beta-, and gamma-cyclodextrins and their derivatives on an in vitro model of blood–brain barrier. J Pharmacol Exp Ther 310:745–751

Nagaraja TN, Knight RA, Croxen RL, Konda KP, Fenstermacher JD (2006) Acute neurovascular unit protection by simvastatin in transient cerebral ischemia. Neurol Res 28:826–830

O'Kane RL, Martínez-López I, DeJoseph MR, Viña JR, Hawkins RA (1999) Na+-dependent glutamate transporters (EAAT1, EAAT2, and EAAT3) of the blood–brain barrier. A mechanism for glutamate removal. J Biol Chem 274:31891–31895

Ponce J, de la Ossa NP, Hurtado O, Millan M, Arenillas JF, Dávalos A, Gasull T (2008) Simvastatin reduces the association of NMDA receptors to lipid rafts: a cholesterol-mediated effect in neuroprotection. Stroke 39:1269–1275

Ramirez C, Tercero I, Pineda A, Burgos JS (2011) Simvastatin is the statin that most efficiently protects against kainate-induced excitotoxicity and memory impairment. J Alzheimers Dis 24:161–174

Robinson MB (2006) Acute regulation of sodium-dependent glutamate transporters: a focus on constitutive and regulated trafficking. Handb Exp Pharmacol 175:251–275

Sierra S, Ramos MC, Molina P, Esteo C, Vázquez JA, Burgos JS (2011) Statins as neuroprotectants: a comparative in vitro study of lipophilicity, blood–brain-barrier penetration, lowering of brain cholesterol, and decrease of neuron cell death. J Alzheimers Dis 23:307–318

Sironi L, Cimino M, Guerrini U et al (2003) Treatment with statins after induction of focal ischemia in rats reduces the extent of brain damage. Arterioscler Thromb Vasc Biol 23:322–327

Sugiura S, Yagita Y, Sasaki T, Todo K, Terasaki Y, Ohyama N, Hori M, Kitagawa K (2007) Postischemic administration of HMG CoA reductase inhibitor inhibits infarct expansion after transient middle cerebral artery occlusion. Brain Res 1181:125–129

Wang Q, Zengin A, Deng C et al (2009) High dose of simvastatin induces hyperlocomotive and anxiolytic-like activities: the association with the up-regulation of NMDA receptor binding in the rat brain. Exp Neurol 216:132–138

Xu Y, Yan J, Zhou P, Li J, Gao H, Xia Y, Wang Q (2012) Neurotransmitter receptors and cognitive dysfunction in Alzheimer's disease and Parkinson's disease. Prog Neurobiol 97:1–13

Yan J, Xu Y, Zhu C et al (2011) Simvastatin prevents dopaminergic neurodegeneration in experimental parkinsonian models: the association with anti-inflammatory responses. PLoS One 6:e20945

Zacco A, Togo J, Spence K, Ellis A, Lloyd D, Furlong S, Piser T (2003) 3-hydroxy-3-methylglutaryl coenzyme A reductase inhibitors protect cortical neurons from excitotoxicity. J Neurosci 23:11104–11111

Zhang L, Zhang ZG, Ding GL et al (2005) Multitargeted effects of statin-enhanced thrombolytic therapy for stroke with recombinant human tissue-type plasminogen activator in the rat. Circulation 112:3486–3494

Appendix

Experimental Procedures

Isolation of Rat Brain Synaptosomes

Wistar rats (males, 100–120 g body weight) were maintained in accordance with the European Guidelines and International Laws and Policies.

The cerebral hemispheres of decapitated animals were rapidly removed and homogenized in ice-cold 0.32 M sucrose, 5 mM HEPES-NaOH, pH 7.4, and 0.2 mM EDTA. The synaptosomes were prepared by differential and Ficoll-400 density gradient centrifugation of rat brain homogenate according to the method of Cotman (1974) with slight modifications. All manipulations were performed at 4 °C. The synaptosomal suspensions were used in experiments during 2–4 h after isolation. The standard salt solution was oxygenated and contained (in mM): NaCl 126; KCl 5; $MgCl_2$ 1.4; NaH_2PO_4 1.0; HEPES 20; pH 7.4; and D-glucose10. The Ca^{2+}-supplemented medium contained 2 mM $CaCl_2$. The Ca^{2+}-free medium contained 1 mM EGTA and no added Ca^{2+}.

Protein concentration was measured as described by Larson et al. (1986).

Extraction of Cholesterol from the Synaptosomes

For the acute treatment of the synaptosomes with M CD (AT), the cholesterol acceptor was added to the synaptosomes suspended in oxygenated standard salt solution and the measurements were performed just after M CD application. For the long-term pretreatment of the synaptosomes with M CD (LP), the synaptosomes were suspended in oxygenated standard salt solution and incubated without (for the control measurements) or with M CD in different concentrations at 37 ° for 35 min. After the pretreatment, the suspension was washed from M CD with 10 volumes of

T. Borisova, *Cholesterol and Presynaptic Glutamate Transport in the Brain*, 69
SpringerBriefs in Neuroscience 12, DOI 10.1007/978-1-4614-7759-4,
© Springer Science+Business Media New York 2013

ice-cold standard salt solution and then centrifuged. The supernatant was completely removed and synaptosomal pellet was resuspended in the appropriate buffer to obtain protein concentration of 2 mg of protein/mL. To answer the question whether the different effects of M CD on glutamate transport in AT and LP were associated with the different duration of M CD treatment and/or with the presence/ absence of the acceptor in the incubating media, we used other methodological protocol of the acceptor application, i.e., long-term treatment (LT). In LT, the synaptosomes were treated with 15 mM M CD at 37 ° for 35 min, however, after the treatment, M CD was not deleted from the incubation media. To determine whether M CD-induced effects were a result of cholesterol depletion, the synaptosomes were also treated with 15 mM M CD complexed with 2.3 mM cholesterol. M CD complexed with cholesterol (15 mM MCD and 2.3 mM cholesterol) was prepared as described by Klein et al. (1995).

The cholesterol content was determined using method described by Findlay and Evans (1987) in aliquots of the untreated (control), M CD-treated, and M CD/ cholesterol complex-treated samples.

Confocal Imaging

The fluorescent probe filipin, which binds to membrane cholesterol (Kruth and Vaughan 1980), was used to clarify the alterations in membrane cholesterol content. The fluorescent dye filipin (50 μg/mL) together with acridine orange, pH-sensitive fluorescent dye (final concentration of 5 μM), was administrated to synaptosomal suspension (final protein concentration of 0.2 mg/mL). Acridine orange was accumulated by synaptic vesicles in the synaptosomes (Zoccarato et al. 1999) and used as additional control of functional state of the synaptosomes. Filipin/acridine orange-labeled synaptosomes were evaluated under the confocal laser scanning microscope LSM 510 META, Carl Zeiss, objective Plan-Apochromat 100×/1.4 Oil DIC, filters: Ch2-1: LP505; Ch3-1: LP650; Ch2-2: LP-420, for filipin 405 nm excitation and >420 nm emission, for acridine orange 488 nm excitation and emission 505–545; >650 nm. For confocal imaging, 1 μL of filipin-labeled synaptosomal suspension (0.2 mg/mL) was squashed and spread between two glass surfaces. Filipin-labeled synaptosomes were viewed in the absence of M CD, then 1 μL of M CD stock solution (300 mM) was added to the thin layer of synaptosomal suspension through the hole in the upper glass at 4 s time point after starting the time series and fluorescence images were captured with camera in each 4.5 s.

Flow Cytometric Analysis of Synaptosomal Preparations

Flow cytometric analysis was performed on flow cytometer COULTER EPICS XL. The synaptosomes were suspended in standard salt solution containing (in mM):

NaCl 126, KCl 5, MgCl$_2$ 1.4, NaH$_2$PO 1.0, HEPES 20, pH 7.4, and D-glucose 10 at a final protein concentration of 20 μg/mL and then incubated at 37 °C for 8 min before the measurements.

The analysis was based on 20,000 particles. All samples were examined with identical instrument settings.

Electron Microscopy of Synaptosomal Preparations

Electron microscopy (H-600, "Hitachi," Japan) of the synaptosomes was performed using the method of ultrathin sections.

Assessment of Lactate Dehydrogenase Leakage

The integrity of the synaptosomal plasma membrane was estimated by monitoring the activity of the cytoplasmic enzyme lactate dehydrogenase (LDH) in the incubation medium. LDH release is recognized measure of disruption of synaptosomal plasma membrane integrity. LDH activity was evaluated spectrophotometrically using the method of Amador et al. (1963), which followed the rate of conversion of NADH to NAD$^+$ at 340 nm. The enzymatic activity was recorded at different time intervals within 0–40 min. Total lysis, corresponding to 100 % of LDH activity on the sample, was determined after treatment with Triton X-100 (1 %). LDH leakage was estimated as a percentage of the total LDH activity in the synaptosomes.

Release Experiments

Intact synaptosomes (for the control experiments and the measurements during AT and LT); cholesterol-deficient synaptosomes preliminary treated with M CD at 37 °C for 35 min followed by its washing (for experiments in LP) and control synaptosomes incubated without M CD at 37 °C for 35 min (for the control in the experiments on LP) were diluted in standard salt solution to 2 mg of protein/mL and after preincubation at 37 °C for 10 min were loaded with L-[^{14}C]glutamate (1 nmol/ mg of protein, 238 mCi/mmol) in Ca^{2+}-supplemented oxygenated standard salt solution for 10 min. After loading, the suspension was washed with 10 volumes of ice-cold oxygenated standard salt solution; the pellet was resuspended in this solution to a final concentration of 1 mg protein/mL and immediately used for release experiments. Release of L-[^{14}C]glutamate from the synaptosomes was measured according to following method: samples (125 μL of the suspension, 0.5 mg of protein/mL) were incubated for different time intervals at 37 °C and rapidly sedimented in a microcentrifuge (20 s at 10,000×g). Release was measured in the aliquots of supernatants (100 μL) by liquid scintillation counting with scintillation cocktail ACS (1.5 mL) and was expressed as a percentage of total amount of radiolabeled neurotransmitter incorporated. Release of the neurotransmitter from the synaptosomes incubated without stimulating agents was used for the assay of tonic release. Stimulated release of the neurotransmitter was calculated by subtracting the basal value from the value of total release.

The experiments with glutamine synthetase blocker MSO are carried out using two protocols: the synaptosomes were preincubated with 1.5 mM MSO for 10 min before L-[^{14}C]glutamate loading procedure or L-[^{14}C]glutamate-loaded synaptosomal suspension was treated with 1.5 mM MSO for 10 min before the release measurements. Both protocols showed the increased extracellular glutamate level in the presence of MSO.

Uptake Experiments

Uptake of L-[^{14}C]glutamate by the synaptosomes was measured as follows: control, M CD-, or M CD/cholesterol-treated samples (125 μL of the suspension, 0.2 mg of protein/mL) were preincubated in standard salt solution at 37 °C for 10 min. Uptake was initiated by the addition of 10 μM L-glutamate supplemented with 420 nM L-[^{14}C]glutamate (0.1 μCi/mL), incubated during different time intervals (1, 2, 5, 10, 20 min) at 37 °C and then rapidly sedimented in a microcentrifuge (20 s at 10,000×g). L-[^{14}C]glutamate uptake was measured as a decrease in radioactivity of the supernatant and an increase in radioactivity of the pellet in aliquots of the supernatant (100 μL) and the pellets by liquid scintillation counting with scintillation cocktail ACS (1.5 mL).

Nonspecific binding of the neurotransmitter was evaluated in cooling samples sedimented immediately after the addition of radiolabeled glutamate.

Amino Acid Analysis

The synaptosomes (15 mL of suspension, 2 mg of protein/mL) were incubated without or with M CD (37 °C, 30 min) in standard oxygenated salt solution, and then each preparation was washed with 10 volumes of ice-cold standard salt solution and sedimented. The pellets were resuspended in 5 mL of standard oxygenated salt solution. For the evaluation of the ambient level of the neuromediator, the synaptosomes (5 mL of suspension, 6 mg of protein/mL) control or treated with M CD were incubated at 37 °C for 15 min, then each preparation was rapidly sedimented in a microcentrifuge (20 s at 10,000×g). Four milliliters of the supernatants was concentrated in rotor evaporator up to 1.5 mL. Two times diluted preparations (3 % sulfosalicylic acid) were analyzed by Amino Acid Analyzer T 339 by the method of ion exchange chromatography.

Vesicle Acidification Measurement

Fluorescence changes were measured using a Hitachi MPF-4 spectrofluorimeter at excitation and emission wavelengths of 490 and 530 nm, respectively (slit bands 5 nm each). The reaction was started by the addition of acridine orange (final concentration of 5 μM) to synaptosomal suspension (0.2 mg/mL final protein concentration) preincubated in a stirred cuvette thermostated at 30 °C for 10 min. The equilibrium level of dye fluorescence was achieved for 15 min. Fluorescence (F) was defined as

$$F = F_t / F_0$$

where F_0 and F_t were the fluorescence intensities of acridine orange in the absence and presence of the synaptosomes, respectively. F_0 was calculated by extrapolation of exponential decay function to $t = 0$.

Measurement of Synaptosomal Plasma Membrane Potential (Em)

Membrane potential measurements were performed using the potentiometric optical dye rhodamine 6G (0.5 μM) that bound with the plasma membrane. The suspension of synaptosomes (0.2 mg/mL of final protein concentration) preincubated at 37 °C for 10 min was added to stirred thermostated cuvette. To estimate changes in the plasma membrane potential the ratio (F) as index of membrane potential was applied:

$$F = F_t / F_0$$

where F_0 and F_t were fluorescence intensities of rhodamine 6G in the absence and presence of the synaptosomes, respectively.

Rhodamine 6G fluorescence measurements were carried using the Hitachi MPF-4 spectrofluorimeter at 528 nm (excitation) and 551 nm (emission) wavelengths (slit bands 5 nm each) (Kasatkina and Borisova 2010).

Synaptic Vesicle Assay

Synaptic vesicles were obtained according to the procedures of De Lorenzo and Freedman (1978) and Roseth et al. (1995) with slight modifications. In brief, the synaptosomes were lysed by rapid resuspension in 1 mM EGTA, 10 mM Tris–HCl, pH 8.1 (3 mL/g of brain tissue) and incubated at 4 °C for 60 min. Preparation was centrifuged at $20,000 \times g$ for 30 min. The supernatant was centrifuged again at $55,000 \times g$ for 60 min and then at $130,000 \times g$ for 60 min. The pellet of synaptic vesicles was suspended in HEPES buffer (10 mM Hepes-Tris, pH 7.4, 120 mM sucrose, 140 mM K-gluconate, 2 mM $MgCl_2$, 2 mM ATP) and immediately used in the experiments. Uptake of L-[^{14}C]glutamate by synaptic vesicles was measured as follows: the samples (125 µL of the suspension, 1 mg of protein/mL) were preincubated in standard salt solution at 30 °C for 10 min. Uptake was initiated by the addition of 50 µM L-glutamate supplemented with 420 nM L-[^{14}C]glutamate (0.1 µCi/mL), incubated at 30 °C for 0–10 min. Uptake was measured in the aliquots of the patterns (100 µL) by filtration through Millipore filters HAWP 025 00 washing with 4 mL warm buffer containing 5 mM Hepes-Tris, pH 7.4, 260 mM sucrose and also in the aliquots of the supernatants (100 µL) and pellets rapidly sedimented in a microcentrifuge (20 s at $10,000 \times g$) by liquid scintillation counting with scintillation cocktail OSC for the filters and ACS for the supernatants and the pellets (1.5 mL).

Nonspecific binding of the neurotransmitter was evaluated in the absence of ATP in the incubation media.

Glutamate Dehydrogenase Assay: The Assessment of the Extracellular Level and Tonic Release of Endogenous Glutamate from Nerve Terminals

The changes in the extracellular level of glutamate in the synaptosomes were detected using glutamate dehydrogenase assay (Nicholls and Sihra 1986; Bezzi et al. 1998). In the presence of glutamate, glutamate dehydrogenase reduced -nicotinamide adenine dinucleotide (NAD$^+$) to NADH, a product that fluoresces, when excited with UV light. The synaptosomes (0.5 mg/mL of final protein concentration) were added to an enzymatic assay solution, which was composed of the standard salt saline, glutamate dehydrogenase (20 U/mL) (Sigma, USA), and NAD$^+$ (1 mM) (Sigma, USA) and preincubated at 37 °C for 10 min. Fluorescence intensity of NADH was measured in a stirred thermostated cuvette (37 °C) at Hitachi MPF-4 spectrofluorimeter at excitation and emission wavelengths of 340 and 460 nm, respectively (slit bands were of 5 nm). Endogenous glutamate released from the synaptosomes to the incubation media was detected as an increase in NADH fluorescence.

To analyze transporter-mediated glutamate release, the synaptosomes were preloaded with cold glutamate (50 µM) at 37 °C for 10 min, then the procedures were similar with those in the experiments with L-[^{14}C]glutamate. The concentration of cold glutamate in the aliquots of the supernatants was determined based on the value of NADH fluorescence in each probe.

In all experiments, glutamate was added to the synaptosomes at the end of the measurements to calibrate the activity of glutamate dehydrogenase.

Statistical Analysis

Results were expressed as mean ± S.E.M. of n independent experiments. Difference between two groups was compared by two-tailed Student's t-test. Differences were considered significant, when $P \leq 0.05$.

References

Amador E, Dorfman LE, Wacker WE (1963) Serum lactic dehydrogenase activity: an analytical assessment of current assays. Clin Chem 12:391–399

Bezzi P, Carmignoto G, Pasti L, Vesce S, Rossi D, Rizzini BL, Pozzan T, Volterra A (1998) Prostaglandins stimulate calcium-dependent glutamate release in astrocytes. Nature 391:281–285

Kasatkina L, Borisova T (2010) Impaired Na$^+$-dependent glutamate uptake in platelets during depolarization of their plasma membrane. Neurochem Int 56:711–719

Cotman CW (1974) Isolation of synaptosomal and synaptic plasma membrane fractions. Meth Enzymol 31:445–452

De Lorenzo RJ, Freedman SD (1978) Calcium dependent neurotransmitter release and protein phosphorylation in synaptic vesicles. Biochem Biophys Res Commun 80:183–192

Findlay J, Evans W (1987) Biological membranes. A practical approach. IRL Press, Oxford/ Washington, pp 103–137

Klein U, Gimpl G, Fahrenholz F (1995) Alteration of the myometrial plasma cholesterol content with -cyclodextrin modulates the binding affinity of the oxytocin receptor. Biochemistry 34:13784–13793

Kruth HS, Vaughan M (1980) Quantification of low density lipoprotein binding and cholesterol accumulation by single human fibroblasts using fluorescence microscopy. J Lipid Res 21:123–130

Larson E, Howlett B, Jagendorf A (1986) Artificial reductant enhancement of the Lowry method for protein determination. Anal Biochem 155:243–248

Nicholls DG, Sihra TS (1986) Synaptosomes possess an exocytotic pool of glutamate. Nature 321:772–773

Roseth S, Fykse EM, Fonnum F (1995) Uptake of L-glutamate into rat brain synaptic vesicles: effect of inhibitors that bind specifically to the glutamate transporter. J Neurochem 65:96–103

Zoccarato F, Cavallini L, Alexandre A (1999) The pH-sensitive dye acridine orange as a tool to monitor exocytosis/endocytosis in synaptosomes. J Neurochem 72:625–633